The Forbidden

Why Families May Never Live Anywhere But Earth

Gerald W. Driggers

Published by Earth-Mars Publishing, Milton, FL
www.earth-marspublishing.com

Printed by CreateSpace in the United States of America

First Edition May 2015

ISBN-13: 978-1511873611
ISBN-10:1511873612

Table of Contents

List of Figures

Introduction

"If no family unit of any type will ever live anywhere but Earth, an age-old question will be answered. Humanity has no place in the universe and will forever more be represented by nothing more than billions of sentient beings on a tiny blue and white ball, endlessly bickering over water and dirt."
Anon.

Historically, families have been relatively easy to identify and describe; they consisted of adults and children. Some complications have been added over the last few decades, but generally, adults have sex and children are born who then repeat the process. Religion, culture, environment, and other factors complicate the details, but basically, that is how the human race has propagated to the point of populating the Earth with over seven billion people.

This process is universally accepted, as it should be; it is simply the way things work. Something related has also shown steady signs of becoming universally accepted. That is the belief that this process will continue when humans venture to other worlds, either in our own solar system or elsewhere. Movies, television shows, books, magazines, and comics have shown people with children wandering about in all manner of spacecraft and on worlds other than Earth.

Nothing could be further from the truth than that scenario. Experience in space cautions that the children of humanity may never be born anywhere but Earth. There are many reasons for this, and we will delve into them as this story unfolds.

Let me state up front that this realization came as a mind jarring shock to me. I spent almost 50 years as a scientist and

Aerospace Engineer (BSAE and MSAE from Auburn University), and then began writing science fiction after I retired. My books (*The Earth-Mars Chronicles*) tell a story of humanity moving adults and children to Mars, followed by children being born there. Those children, spawned in my imagination, were different and superior relative to their cousins on Earth.

One day as I pondered the next part of the series, which would introduce the third and fourth generations, it occurred to me that there was no scientific basis for my optimism. I had not been able to find a single substantiated basis for assuming that the evolutionary changes would be positive. Oh sure, I wrote good words about the specialists of the future being able to prevent the bad things from happening thus making sure that only good changes took place, but the seed of concern had been planted. I began to dig deeper into the state of knowledge surrounding current experience and research related to human physiological changes in the presence of different states of gravity.

Here is a brief summary of what I found.

1. People who go into orbit about the Earth and exist in "microgravity" (zero gravity as far as the human body is concerned) begin to experience substantial physical degradation almost immediately. The degradation continues for as long as the individual is in space.

2. Drugs and hours of strenuous exercise each day are required to make sure return to Earth and recovery are possible.

3. No data exist concerning what will happen to astronauts on Mars for an extended period of time where the gravity is 38% of that on the Earth.

4. No one even dares to speculate as to whether children can be conceived, develop in the mother's womb, be born, and mature normally in any conditions other than those on Earth.

There are many more details behind each of these statements, some of which will be discussed in the following chapters. First, however, I wish to set the stage a bit with regard to why many of us care about this and why we as a species need to find answers as to whether human families will be the forbidden, denied forever any expansion into the heavens. Finally, I will present one plan (I am sure there will be more) which can provide the answers we need.

Why Now?

It has been fifty years since the United States had a destination and goals in space. Fifty years ago (half a century!) the Moon was that destination; the goals were beating the Soviets and returning safely. Now, finally, there seems to be a growing acceptance that Mars will be our next destination.

However, opinions are split on what the goals should be. One camp calls for a safe return, the other advocates immediately establishing a settlement after a successful one-way trip. This divergence of goals creates two separate but overlapping sets of technical, physiological, and psychological requirements. Both approaches share critical issues with regard to astronaut health and viability that are precipitated by gravity, radiation, and the environment on Mars.

Mars is far away; many times farther than the Moon and, unlike our lunar companion, the separation between Earth and Mars is constantly changing. The Apollo astronauts had abort-to-Earth options for emergencies. The trip was measured in days. Astronauts traveling to Mars will not have such an option; their return to Earth will be measured in months and years.

The separation of Mars and Earth is so great that it takes electromagnetic waves (light, radio, television, radar, etc.) several minutes to travel from one to the other. That is not seconds; it is minutes as in about six at best (one way) and 22 at greatest separation. Those who devote their days and nights to controlling and driving rovers on Mars (Curiosity for almost three years and Opportunity for almost 12 years at this writing) know firsthand how challenging that can be.

However, there are other much more critical issues to confront. Particularly distressing is that gravity on Mars is 0.38 (38%) of what we experience all of our lives on Earth. As will be discussed in some detail later, changing a human's gravity environment sets physiological changes in motion immediately. In addition to the adaptive changes within humans, there will be equipment with processes essential to survival on Mars that could have some sensitivity to changes in the level of gravity, and their functionality will need to be determined with certainty. Data on human response to 0.38 gravity does not exist, but there is one thing that we already know absolutely. Many human body functions are extremely sensitive to a *total loss* of gravitational forces.

For obvious reasons, experiments concerning extended exposure at less than one g cannot be conducted on Earth, and until recently, priorities on the International Space Station (ISS) have negated attempts to install experimental centrifuges. A very small one was recently installed for use with insects and rodents.

Most of the data on animals in artificial gravity of less than 1g has been limited to work by scientists of the former Soviet Union. Soviet scientists did experiments onboard satellites in the 1970's that suggest that as little as 0.3g works well for maintaining turtle and rodent muscles, but it cannot be considered conclusive for humans.

A grand life-size experiment in space inside of a centrifuge creating simulated Mars gravity is the only way to find the answers required. There is no question that NASA, or whatever entity ultimately builds such a facility, will give it a catchy name derived from a snappy acronym. For purposes of discussion, I have named it the Mars Close to Home (MCTH) facility or station.

This concept is far from being a modern original. Design ideas for stations in space that rotated to create artificial gravity have been around since the beginning of the last century. Only recently has the need for one creating less than one Earth gravity (1g) begun to surface. The major near-term issues are: (1) How much artificial gravity (AG) is needed to keep astronauts healthy during long space transits; and (2) will those landing on Mars be required to follow a strict regimen of drugs and strenuous exercise in order to return home capable of readapting to Earth gravity.

Such a facility is also essential to answering conclusively questions that could seal the fate of settling Mars forever, which will effectively relegate human families to Earth. These critical issues may negate ever having permanent settlements on Mars. First, will the human fetus develop properly in the womb and deliver normally? The second issue is how will growing up in 0.38g affect the development of a child born on Mars? Answering these questions will require substantial facilities and many years of research possibly coupled with extensive debates over ethics. Proper research facilities in Earth orbit must be designed to accommodate such work.

Why Mars?

Mars is very hostile toward any form of life from Earth including humans. However, it is the only inhabitable planet in our solar system other than Earth. Some of you may logically ask, "Why not just go to a nicer planet in another solar system? We should have a warp drive or some other propulsion system capable of doing that in a few years." (Spoiler Alert!) Wrong!

I will not bore you with the details (there are plenty of books out there prepared to do that), but all of those television and movie interstellar travel schemes are just so much pure mumbo-jumbo fantasy. Actually, the more we learn about matter, energy, and the universe, the more *unlikely* interstellar travel becomes. The solar system is all we have now, and it may be all that the human race will ever explore in person, but that is a discussion for another time.

In other words, given what we know right now, it will have to be Mars. Granted, it would be more interesting if there were a lush climate and an alien civilization waiting for us to visit, but there is not. Mars at first glance appears pretty desolate, but it is much more interesting than some would have you believe. However, the most intriguing thing about Mars is that it offers the human race an opportunity to build a haven for our species and our modern civilization while exploring an entire planet.

It offers safe haven because the remoteness and the relatively benign nature of the Red Planet isolate it from the various phenomena of nature and human activity that threaten to end civilization as we know it on Earth. Even a large asteroid capable of wreaking havoc on Earth would have only minor effects on the settlements of Mars as long as it did not hit too close. That is one circumstance where lack of a substantial atmosphere and oceans is a good thing!

There are many more good reasons to establish humanity on Mars that will be further explored later. However, it is not the intent of this work to convince the world at large to back an exploratory visit to Mars. What I will present here is a wakeup call to the fact that any vision of forming permanent settlements on Mars where women have children and families expand into multiple generations *is in serious jeopardy and may, in fact, be impossible. If so, then humanity may never become a multi-planet species.* If that is the case, then we should turn our attention to building the large in-space habitats envisioned by Dr. Gerard O'Neill during the latter part of the last century.

Mars has been receiving substantial (comparatively speaking) scrutiny in the national and world press recently, so let us dig into why, followed by a summary of why the Red Planet should be receiving even more attention in the next few years.

Generations of Dreamers

Fear not! This chapter will not bore you with long drawn out discussions detailing the wonderful visions for Mars that have occupied the thoughts of many great men and women of science and literature. Everything you wish to know on that subject is available through any search engine after a few insightful stabs at the keyboard. These brief paragraphs are addressed to the current and future generations of dreamers.

Two generations of dreamers put humans on the Moon in July of 1969. That feat becomes more remarkable when noting that Robert Goddard launched the first successful liquid fuel rocket in the U.S. in 1926, only _43 years_ before Apollo 11 touched down. It is equally astounding to note that only _35 years_ separated the launch of the first two successful liquid fuel rockets by amateurs in Germany (enter Werner von Braun) and the first landing on the Moon. Only 35 years; half my age.

Dreamers were largely responsible for the achievements of the U.S. space program of the sixties, but, contrary to the notions of some commentators, they did not _cause_ it to happen. Politics, both national and international, motivated President Kennedy's monumental decision. Do not take my word for it; stroke your computer if you want more. Millions of words have been written on the subject.

If the dreamers of the sixties and seventies had been able to _cause_ things to happen, another 35 years would have seen humans land on Mars and set up permanent camp. By now (46 years after Apollo 11), there would be mines on the Moon, a thriving Cis-lunar trade, and settlements growing on Mars (if such things are feasible). The problems, challenges, and hang-ups that people agonize over now regarding these objectives would all be old news, and we would all be reaping the benefits.

***

But I digress. Let us put some things in perspective. Break out a quarter and a dime, or some currency of similar dimensions. A quarter is 25 millimeters in diameter, slightly less than one inch. Go to your local American football field and place the quarter next to one goal post. Amble 360 feet (120 yards, 110 meters) to the opposite goal post and lay the dime on the turf. That is to scale the closest that the two planets *ever* come to each other, and that occurrence is rare. By the way, you would have to have a microscope to see a spacecraft suitable for the trip to Mars at that scale. It would be 1/10,000 of a millimeter long! When the two planets are at their greatest separation, the quarter and dime would be 7.36 times further apart (one-half mile or 810 meters).

History shows that the nation solving all of the problems, challenges, and hang-ups of such an endeavor will reap a pile of benefits in technology advancement, new knowledge, unique capabilities, economic growth, and international prestige. The only problem is, those benefits are a real bear to predict, and we have an econo-politic system that demands feel good, sure-fire predictions.

This is not to say that a few of the technology and hardware developments that are on the critical path to traveling to Mars and remaining there are not being funded and worked on. NASA and other space agencies, both in the U.S. and other countries, have active R&D programs addressing *some* of the critical issues. The problem is that

16

technology efforts based on nebulous "plans" are often devoid of focus, continuity, appropriate requirements, and timing.

One area that is widely accepted as a complete unknown is how the human body will react to the 38% Earth gravity (0.38g) of Mars. Our bodies have evolved to be finely tuned to operate in the gravity field of Earth, and it reacts rapidly and decisively to being in micro-gravity, such as in orbit around the Earth. Immediately upon experiencing micro-g, the human body begins to retune things. Muscles begin to atrophy, bone density begins to decrease, blood redistributes itself, fluid pressure in various organs change, and a number of degenerative states begin to manifest throughout the body.

The stay on the Moon by the Apollo crew was too short to get any definitive health indicators to compare to their counterparts on the ground or later in the ISS. Therefore, the scientists and space medicine health experts have no basis for extrapolation to any partial gravity conditions. That covers Mercury, the Moon, and Mars for the inner solar system. Venus has a gravity field with very near the same attraction as the Earth, but the extremely hellish surface conditions make it an unlikely candidate for visitation, much less settlement.

Serious (i.e. medical) concerns about the effects of "free-fall" or micro-gravity have been around since the early part of the last century when pioneers such as Tsiolkovsky, Oberth, and Goddard began to seriously consider the real possibility of travel to and through space. The modern response has been to address by pharmacology and physical exercise the issues that surfaced during Mercury, Gemini, Apollo, Skylab, and ISS, as well as all of the Russian human space flights.

Proposals to use artificial gravity through rotation-induced centripetal acceleration as a remedy have been largely ignored. This was not due to a lack of knowledge or interest from researchers. The proposals to address the issue of reduced gravity along with microgravity were simply ignored or cut from the program due to higher priorities. The long-term significance of these shortsighted decisions will become more obvious later.

What Does "Settling Mars" Mean?

Excellent query, and there is a corollary question. What does "exploring Mars" mean? We will get to that a little later. First, let us dwell on a definition of "settling Mars" since that understanding will influence all further discussion.

If someone has established a standard definition for "settling Mars," I have not been able to locate it, so I consulted Google for "define settle":

make one's permanent home somewhere.

synonyms: make one's home in, set up home in, take up residence in, put down roots in, establish oneself in;

establish a colony in.

synonyms: colonize, occupy, inhabit, people, populate

Of the definitions offered, these seemed to fit rather well. We are discussing people (human beings) leaving Earth and putting down roots in one or more locations on Mars, and we do intend to populate the planet. So, here is my working definition of "settling Mars."

Settling Mars: The act of human beings willfully establishing permanent residence on Mars, remaining healthy and content, and raising viable offspring to maturity, followed by an indefinite number of subsequent equally viable and content generations.

Again, from Google:

content:

in a state of peaceful happiness.

synonyms: contented, satisfied, pleased, gratified, fulfilled, happy, cheerful, glad

This is a somewhat restrictive definition, since it requires that healthy children be born and grow to viable maturity and subsequently repeat the process ad infinitum. Also, observe that this definition of settling Mars has long-term overarching requirements. The Settlers must, by and large, have contented lives with said state extending into the future generations ad infinitum.

Subsumed in this definition are thousands of details related to who the Settlers are, how they get to Mars, where they live, what kind of life they have, their jobs, health care, education, pay, and on-and-on. Also embraced by this definition are key failure modes such as, living on Mars cannot make Settlers or their offspring repeatedly or permanently unhealthy.

This definition does not require that those humans who leave Earth and settle on Mars or their offspring bear any particular resemblance to those who remain on the home planet. Facing and accepting the evolution that will take place for those humans born and growing up on Mars may turn out to be one of the most difficult challenges.

Exploring Mars

Now to the definition of "exploring Mars." Again, there is not a generally accepted definition that I have been able to locate. The authors of most articles, videos, and books use the terms "exploring Mars" and "Mars exploration" to cover every activity that returns a new image or piece of information. I find this less than fulfilling. It is a little like someone landing in the Mojave Desert and proclaiming that they are "exploring Earth!" Okay, the statement is technically valid, but it is grossly misleading.

All of the flybys, orbiters, and landers combined, whether stationary or "roving," have not "explored Mars," and if we send 100 more, they will not have "explored Mars." The planet has a surface area comparable to that covered by all of the continents on Earth. Additionally, none of the past landers visited the most interesting areas on Mars, and they may never do so. Why? Because the landing area and surrounding terrain will be considered too dangerous.

This is not to imply that those areas available for safe landings are not of interest. The exploration executed by the Mars Science Lab on the rover Curiosity in Gale Crater proves that conclusively. That, however, does not diminish the fact that robotic exploration of certain large portions of Mars is probably not practical.

Now that I have the robot-versus-human debate inflamed, it is appropriate to point out that the subject will not be examined here. Suffice it to say that "exploring" Mars is not the same as settling Mars, and I am sure that every reader already knows that. The time spent on the subject here is useful because there are very close bonds between carefully planned exploration and establishing settlements, and because exploration will play a major role on the path to answering a fundamental question. Is the vision of settling Mars, based on my definition, an impossible dream, or a reasonable expectation?

Oh, as far as a definition for exploring Mars, let us use "any activity that increases humanities knowledge of the Red Planet." Now it is time to get into some nitty-gritty details related to settling Mars.

You Want to Live Where?

Parents, family, friends, acquaintances, and complete strangers are asking a few thousand people this question. They are the individuals from around the world that made application to Mars One, a non-profit organization in the Netherlands, to become Martians. Mars One, contrary to what you may read or hear, is making a whole-hearted attempt to place four people on Mars before 2030 (their target as of this writing) where the chosen few will begin a settlement and spend the rest of their lives without possibility of returning to Earth.

These intrepid souls, under the Mars One plan, will be joined by four more heroes (or nut cases, or both, depending on your point of view) every two years when the planets align properly. If you are interested in more details of their plans you will find it (albeit rather high level) at www.mars-one.com.

Mars One was founded in 2012 by Bas Lansdorp and Arno Wielders of the Netherlands based on their uncompromising belief that humanity and our technology are postured to take the next bold step in establishing the future of the species. That step, according to Mars One and their followers, is for the human race to become a multi-planet species. Simple to say, but, as we shall see, it is incredibly difficult to do. As we all know, attaining such heights of achievement requires extraordinary people, tremendous individual effort, and huge sums of money. This characterization applies well to settling Mars.

There is a little bit of a hang-up in the Mars One concept, however. From the Frequently Asked Questions section of the Mars One website:

"Can the astronauts have children on Mars?"

"Mars One will advise the first settlement inhabitants not to attempt to have children because:

- *In the first years, the Mars settlement is not a suitable place for children to live. The medical facilities will be limited and the group is too small.*

- *The human ability to conceive in reduced gravity is not known, neither is there enough research on whether a fetus can grow normally under these circumstances.*

In order to establish a true settlement on Mars, Mars One recognizes having children is vital. Therefore this will be an important point of research."

Take special note of the phrases *"having children is vital"* and *"an important point of research"*; more on these topics later. First, here is a short background summary on the subject of settling Mars.

Contrary to what many seem to believe, the concept of visiting and possibly even living on Mars is far from new. Debates about life there and visiting our neighbor began soon after the discovery of the planet. Speculation became rampant after Italian astronomer Giovanni Schiaparelli declared in 1877 that he had observed "canali" (actually translated to "channels," but who wanted to hear that?) on Mars. Canals equated to an intelligent, industrious, and widespread population of beings on Mars. Works of fiction and non-fiction speculated and provided inspiration for the next 80+ years.

That is why I am so enthralled by the prospect of settling Mars and now find myself writing books about it. The "space bug" bit me when I was eight or nine years old. It was not just Mars that captured my imagination; the planets were mysterious worlds about which little of a detailed nature was known. There was considerable speculation in the middle of the last century about beings that might live on Mars or Venus. The harshness of the environments on those planets was becoming known to scientists but that did little to curb the imagination of a certain young boy (and many grownups) in the 1950s. Figure 1 captures the imagination of the 1950s. (Credits for Figures are found on the List of Figures.)

After Sputnik was launched in October 1957, most inhabitants of the small town where I lived were immediately interested in and afraid of rockets and

Figure 1. Mars was still a fantasy world in 1958

space travel. My strange interest in space was suddenly newsworthy. Of course, a couple of years later I discovered girls, beer, and drag racing and my interests took a more mainstream point of the compass, but I never forgot my dream.

However, all hope for intelligent or higher-order life forms on Mars was pretty much dashed by the images and atmospheric data returned from Mariner IV on 15 July 1965. Author George R. R. Martin expressed it well in his Introduction to the excellent short story collection *Old Mars*. "It was Mariner that put an end to the glory days of Old Mars…"

The images and atmospheric data from Mariner IV almost devastated me; Mars was just like the Moon! Later I would come to appreciate that nothing was farther from the truth, and my relationship with Mars warmed again.

Unfortunately, that was just in time for all of those dreams of exploration and settlement to be postponed indefinitely in favor of a fantastic technological achievement and political compromise dead-end program called the Space Shuttle.

Even though Mars as an objective for human exploration disappeared into the background noise of everyday life, my fascination with the "real Mars" continued to grow. Today I experience the same intrigue that took me through time and space as a lad. Part of the reason for my change in attitude toward the "real Mars" has been a growing appreciation for two things.

First, Mars turns out to be a fascinating planet in its own right. Sure, there are no intelligent and mysterious aliens with cities and cultures to make contact with, and you have to walk around in a space suit, but so what? Here is an entire world whose evolutionary history has been written with its own stylus. With roughly the same surface area as the landmasses of Earth, Mars represents lifetimes of exploration and generations of new knowledge, not to mention an opportunity to participate in one of the greatest adventures in human history!

Second, Mars is the only planet within human reach that we can reasonably land on and inhabit. Of course, we could go to certain twilight areas on Mercury and futz around but so what? That is pretty much all we would do there. Floating around in the upper atmosphere of Venus has been proposed, but few people seem inspired by the concept.

Mars, on the other hand, has decent gravity (38% of Earth); a similar rotation rate to Earth (a day is 24 hours 39 minutes 35 seconds); harsh but manageable climate; enough atmosphere to be meaningful; and a plethora of available resources. For many excellent discussions of settling Mars vs. the Moon, I direct you back to your search engine. Dr. Robert Zubrin, founder and President of The Mars Society, is particularly eloquent on the subject.

I have spent untold hours exploring the Red Planet through the magic of Google Mars, and I have only "scratched the surface" so to speak. Always I long for higher resolution and the ability to drill, dig, move, look under, see more closely, crack open, go inside of, etc. In the final analysis, I believe Mars cannot be explored from Earth or from orbit. In addition, robots will be very useful for many things, but they will never replace humans when it comes to exploration. That statement will be the subject of debate long after the exploration of Mars has truly begun, but that is my opinion.

Lest the reader misunderstand, let me reiterate that exploration is only one facet of settling Mars. Exploration will provide both an incentive and a source of income in the settlement equation, but it is doubtful that it engenders sufficient motivation for such a major undertaking as large-scale settlement. (Exploration is sufficient motivation for me, but unfortunately, I cannot write credible checks that large!)

So, is settling Mars an impossible dream, just a long shot, or a reasonable expectation? I hope it will turn out to be the latter and that this book will bring insight and fortification against the doubts that at times, threaten to conquer us all. However, there are substantial obstacles to the long-term settlement of Mars, and they will be discussed in the following pages. The most vexing are the questions related to procreation, childhood, and evolution of the human species in the environment of Mars.

People Will Immigrate to Mars Because ...

Each astronaut application to Mars One includes a statement in the person's own words about what moved them to apply. The responses were, in general, quite predictable. Historically, science fiction has been full of reasons for heroes and villains to go to Mars. Recent history on Earth, especially the settling of new frontiers, also presents us with a number of possible motivations for humans to undertake such an audacious and perilous endeavor.

What follows is an informal compilation from these sources presented in no particular order. There are certainly many more, but this list leads to some interesting questions that will need to be addressed at some point in the settlement process, some sooner than others.

Keep in mind that all of this is sheer fantasy unless the cost of transportation, room, board, and the ability to afford the trip converge. If we based our discussion on the historical cost of space programs, there would be no discussion. As already noted, there are reasons grounded in health and evolution that may also prevent the settlement of Mars. Should any of these come to fruition, the human race might be doomed to whatever terminal fate awaits us on Earth, and genuine science fiction will be limited to depressing dystopian plotlines. That is, unless we build the O'Neill colonies.

Well, that scenario is no fun for us Mars heads, so let us wish Elon Musk, Mars One, and the other activists good luck and forge ahead! We will return to the obstacles later. Right now, I am hoping to make a convincing argument that the problems are worth solving because a whole lot of people really do wish to emigrate to Mars.

Here is my list. Each one is elaborated on below.

Why will people choose to settle on Mars?

1. Go as explorer and decide to stay.
2. They harbor unrealistic expectations.
3. To find a job.
4. Seeking an opportunity to become "rich" on Earth, on Mars, or both.
5. Because they do not have a choice.
    a. Exiled felons
    b. Life on most or all of Earth begins to become unbearable
    c. Totalitarian government selects and orders people to Mars
6. Adventure
    a. Precarious lifestyle
    b. Exploration
7. To make history and be remembered
8. To escape something on Earth
    a. Incarceration
    b. A rich overbearing family
    c. Poverty
    d. Responsibilities
9. To be a pioneer
10. Enjoy more personal freedom compared to their nation of origin
11. Ensure that their children have opportunities for a positive future
12. Paid high salary to go
13. Given property plus in-situ support to expand agriculture within a settlement or an environ
    (See the definition of "environ" below.)

14. Given property including mineral rights and local support in exchange for expanded mining

15. An artist (of any medium) seeking new inspirations (I know this is a *real* long shot!)

16. To retire. I have to include this because Elon Musk brought it up in an interview. Honestly, I do not believe I would ever have come up with it on my own.

So that is my sweet sixteen. Obviously, the cost of transportation and cost of living (sustainment expense) would figure heavily into determining which if any of these would ever be practical. That little problem will be discussed further in later chapters. For now, I must assume that the price point for transport, lodging, and board will fall within the capability of some individuals, groups, governments, or institutions described below.

**

The following are comments related to each of these potential motivations. At the end of each brief discussion are comments on where funding for the move might come from. Treat each one as the result of brainstorming, not as if derived from extensive research!

1. Go as explorers and decide to stay.

Why would they do this? Obviously, one or more of the other 15 reasons could apply, unless of course they just fall in love with the planet or a Settler and cannot bear to leave. Actually, I have seen one other reason presented in multiple works of fiction, and it raises its own issues. In those stories, a female astronaut gets pregnant and has to stay to raise the infant. That is one way to get an ironclad guarantee of future supply missions, assuming you can last for two-to-four years until they get there, the baby survives, etc., etc. This scenario also assumes that fetus and child will develop in a healthy fashion. Unfortunately, there is no evidence, anecdotal or otherwise, supporting that assumption.

An explorer might decide to stay for a number of reasons once the settlement process has a foothold. They would have valuable skills and might be offered attractive reimbursement or particularly interesting work, such as conducting exploration in exchange for a fee. This sounds far-fetched from our present perspective, but when taken in a broader context it may not be so outlandish. An open mind is essential to keeping a dream alive!

Financing – The exploration program will have paid for his or her trip to Mars.

2. They have unrealistic expectations.

These are commonly considered to be the "Space Cadets" and utopia hunters, seeking unlimited freedom to do whatever they please, live a fantasy lifestyle, get rich quick, or similar gratification. It is difficult to see how anyone with unrealistic expectations would be in the program for long once the screening process begins, but perhaps if they are paying their own way they might slip through. This is a category of potential Settler that will receive no further assessment.

Financing – N/A

3. To find a job.

Robert Zubrin and others have written about possible economic futures for Mars settlements and they have pointed out that migration to find employment has been a powerful motivator throughout history. Obviously, taking it to Mars is a different ballgame, but it does seem reasonable to assume that there will always be a labor shortage as settlements grow and new ones are built.

Historically, frontier settlement has been a labor-intensive enterprise, frequently making use of local labor. Unfortunately, such occurrences were often under less than honorable circumstances. However, historical trends do not apply well here. One very good reason will be the robots, which are one factor making many things difficult to prognosticate. The division of labor between humans and robots may be determined by such things as radiation exposure times, job difficulty in a surface suit, inherent danger, etc.

We honestly cannot forecast the human and robot labor split, but there are a few qualitative things to consider. First, if our objective is to build a long-term viable branch of the human species on Mars, why would robots be given any advantage in the job market? Second, robots will not be in management positions. Okay, I admit that is subjective and only time will tell. Third, robots cannot build people; they can only build more robots.

Financing – government subsidy, family pools resources, loans, indentured, hired to go to Mars

4. Seeking an opportunity to become "rich" quick on Earth, on Mars, or both

Some prime motivations would be greed, ambition, deep-seated desire to "show someone," and others.

Financing – inheritance, loans, investors, savings, pooled resources

5. Because they do not have a choice

a. Exiled felons – Will not happen, because the general population on Mars will not accept it. The risk would be totally unacceptable.

Financing – N/A

b.   Life on most or all of Earth begins to become unbearable – The wealthy would buy up most or all of the seats available to go to Mars. This is a self-limiting scenario because at some point the supporting infrastructure making Mars flights and supplies possible would break down, and flights would stop. Not a very jolly scenario (but still the backbone of one of my books)!

Financing – Wealthy self-fund, government

c.   Totalitarian government selects and orders people to Mars – The government motivation that would drive such an act is difficult to fathom unless they need some specialties on Mars that could not be filled through volunteer recruitment. Seems pretty far-fetched, but not impossible.

Financing – government

6. Adventure (This is a very broad category. I only picked two example sub-motivations.)

a.   Precarious lifestyle - Although accused of being given to a wee bit of this myself over the years, I do not fully understand why people embrace it as a lifestyle. However, they are out there, and some have significant resources.

Financing – self, family, loans

b.   Exploration – Here is a motivation that most people can sink their teeth into. Historically, it is the tried-and-true way for an adventurer to quickly acquire recognition and a place in the annals of history.

Financing – self, family, government, institutions, entertainment industry, sponsors, others

7. To make history and be remembered

This motivation is not the same as number 6, Adventure. President Lincoln pushed and browbeat the Emancipation Proclamation through the Congress not just because he saw it as the right thing to do. He wanted it to be a part of his legacy, something he would be remembered for along with the Civil War. There was little in the way of adventure in his motivation. Many of the early Settlers on Mars will be seeking to be remembered in history and lore.

Financing – self, entertainment industry, institutions

8. To escape something on Earth

a. Incarceration – It is doubtful that someone will be able to spoof future recognition techniques. This is also a motivation that is not very interesting to us in light of the context of this document.

Financing – N/A

b. A rich overbearing family – This has happened before. Just consider how "away" one can get by going all of the way to Mars!

Financing – self, family

c. Poverty – My first impulse is to dismiss this motivation out of hand as intuitively unrealistic. However, people have immigrated before when it seemed impossible to accomplish by pooling resources and then boot strapping, or by working out an indenture.

Financing – pooling resources, indenture, loans

d. Responsibilities – This is another motivation that is difficult to identify with, but many a sailor has gone to sea (to pick one example) seeking escape from responsibilities. Mars could be the ultimate escape!

Financing – self, indenture, loans, family

9. To be a pioneer

These people are motivated by what is commonly labeled the "pioneering spirit." They are represented on Earth by those who choose to live "off the grid" and be as independent as possible, often living in circumstances most people would categorize as somewhat primitive.

Financing – self, like-minded group anticipating meeting up later on Mars

10. Enjoy more personal freedom compared to their country of origin

This motivation brought and continues to bring many people to the United States and other nations, but why Mars? Why not seek freedom in a nation on Earth?

Financing – N/A

11. Ensure that their children have opportunities for a positive future

Responsible parents have always had this motivation. A number of the previous motivators could couple with this one.

Financing – Self, family, group, indentured

12. Paid high salary to go

This is perhaps the simplest of motivations that people yield to. Remote jobs all over the Earth from the North Sea to Antarctica are filled by people lured by large salaries. The catch with Mars of course, is that it might be a one-way trip if they go there as a Settler.

Financing – Self, family, company, government

13. Given property plus in-situ support to expand agriculture within a settlement or an environ

This seems a little far-fetched, but it might occur in a settlement with sufficient labor shortages or a lack of interest in farming. Such actions have been taken by several nations on Earth in the past. (An "environ" in my later novels is an alternative to terraforming popularly called paraterraforming. An environ consists of an airtight enclosure large enough to encompass one or more settlements and provide a "shirt sleeve" environment.)

Financing – settlement, Earth government

14. Given property including mineral rights and local support in exchange for expanded mining

This also seems a little unlikely, but it might occur in a settlement with sufficient labor shortages or a lack of expertise or interest in mining. Such actions have been taken by nations on Earth in the past when resource availability came into question.

Financing – settlement, Earth government, Earth companies

15. An artist (of any medium) seeking new inspirations (I know this is a *real* long shot!)

Financing – self, patron, sponsors

16. To retire. I have to include this one because Elon Musk brought it up. Honestly, I do not believe that I would ever have thought of it on my own.

Financing – Self

***

"Why go through all of this?" you might ask. The answer is simple. This discussion is a response to the "giggle factor." Human settlement of Mars is not a concept in the world of fantasy. It is a subject worthy of time investment and serious thought. If the human race cannot evolve in an acceptable fashion on Mars then we have no other candidate planet in our solar system. This also includes our Moon where the attraction of gravity is only 0.17 (17%) of that on Earth.

Another reason for looking at potential futures and inhabitants on Mars is to give some credibility to the currently rather alien (to some people) concept that humans would actually be interested in emigrating to Mars. The point is that we, as human beings, are poised to lose something of immeasurable consequence if the environment (especially gravity) on Mars dictates that we cannot live as healthy and happy families anywhere but on Earth.

A very positive thing about this list is the breadth of it. Many skills and education levels fit within these categories. Also, white collar, blue collar, wealthy, poor, highly educated, moderately educated, and many more categories of humanity are included.

To complete the point, let us look at what kinds of business and industry we could expect to be on Mars.

Business and Industry on Mars

With good foresight and planning, each settlement on Mars would have synergistic specialties that, when taken as a whole, would make the totality of settlements very nearly self-sufficient. Each settlement can be expected to do independent farming sufficient to provide for their own needs but all can be expected to have specialties with which to trade, barter, or sell.

I have not seen a rigorous analysis of what level of industrialization, automation, and population will be required to achieve self-sufficiency; one needs to be done when enough information is available to support it. We know intuitively that it will not be necessary to duplicate industries covering all categories of the Standard Industrial Classification (SIC) and North American Industry Classification System (NAICS). Likewise, it is safe to assume that fabrication capabilities transplanted to Mars will be optimized for flexibility and re-configurability. Therefore, one machine and operator will cover an entire population of SICs and NAICS. However, self-sufficiency may be delayed indefinitely if that capability is not woven into the fabric and foundation of settlement planning.

Trade between Settlers will probably occur relatively early in the settlement process. That is human nature and the way of our ancestors all over the Earth. Trade between settlements would evolve as specialties develop and eventually, there would be trade with entities on Earth. We can only speak in generalities here, but some speculation on goods and services seems reasonably sane (or at least fun).

1. Any type of gems from Mars can be expected to have value on Earth well above diamonds or any other precious stone, especially if they have some feature that uniquely identifies their origin as the Red Planet. Impactite is a term used to describe any rock that was formed due to melting during asteroid or meteorite impact. Some of these are much rarer than diamonds (which are also found in impact craters). Martian meteorites found in Antarctica, which are thought to represent samples of the Martian crust, have HSE (Highly Siderophile Elements) values that resemble groups of iron meteorites and stony irons, suggesting that a process of raw materials deposition by meteors and asteroids similar to that which took place on Earth also took place on Mars. The exploration process should include characterizations specifically designed to identify and quantify the population of impactites in selected craters.

2. If the recent and current decorating trends by the well off in developed nations continue, items of any type (stones, sand, minerals, etc.) from Mars should have great value if marketed properly on Earth.

3. A major product delivered utilizing electronic export would be the results of specific explorations. Clients on Earth (government agencies, universities, companies, individuals) would pay for exploration results. The knowledge gained would be had at far less expense than sending a highly sophisticated robot from Earth to conduct the same exploration, and the risk of failure is almost non-existent.

The search for present or past life will continue for a long time. If building blocks or other hints of life were found in the future, there would be a substantial desire for further data and evidence, specimens in particular.

4. Entertainment from Mars will have audiences on Earth. The reduced gravity would lead to unique dance and acrobatics of interest to many. Competitive sports adapted to Mars or unique to the Settlers would also generate substantial interest. Drama stories shot on location on Mars should find a niche in the world of entertainment. Undoubtedly, there could be many more opportunities in the "show business" arena.

5. Scientific knowledge gained in the unique environments of Martian laboratories would draw sponsors. In addition, unpredictable but unique products can be expected to come out of the laboratories with markets on Mars and Earth.

6. If transportation costs are low enough, it would make sense to mine rare elements on Mars and sell them to clients on Earth. The minerals that lie below and near the craters on Mars are the product of horrendous forces and environments that are extremely difficult if not impossible to duplicate in a laboratory or factory. No one would know what could be found until someone starts digging (metaphorically speaking). They would sure know where to start looking, though. The challenge would be selecting the craters to dig in and near. One thing about it; the Settlers would not have to go to the asteroids for anything. The asteroids already came to them!

7. Tourism would have a good long-term potential to become a viable enterprise if three things happen. First, the travel time would have to be shortened a lot; second, the 26 month wait to return must disappear; and third, the world's economy must remain sufficiently vigorous to support a large cadre of wealthy capable of paying the fare, room, and board. This of course assumes that transportation costs will come down to a level that favors extravagant vacations or other exclusive jaunts.

***

Clients on Earth would pay for these products and services, probably by depositing the agreed to reimbursement in specified funds on Earth. How these funds would be used by the Martian entity receiving them would depend on the specific nature of the economic system in effect on Mars. It is reasonable to expect that a free enterprise system would hold sway in the mature settlement environment. There will always be entrepreneurs who can figure out how to further their compensation and provide the best for their family. There is no reason to expect this to change on Mars.

As with true international trade on Earth, (not including "assembled elsewhere" type products) interplanetary commerce between Earth and Mars would consist of unique goods not readily available on the other planet. In the collective vision of we Mars dreamers, there would be slow ships (solar sails, ion propelled craft, others) for trade that was not time critical, and high performance ships (Nuclear Thermal, possibly some chemical, others) for passengers and high value products or emergency supplies.

It is not difficult to imagine that certain spirits, wines, cultural décor, collectibles, very specialized electronics, frozen foods, and numerous other items would be in demand by the Settlers who have credits to spend on Earth in exchange for products and transportation. Of course, there would be electronic exchange items of trade as well. Entertainment and special programming broadcasts from Earth fit into this category, as well as most of the exploration results on Mars discussed earlier. The results of research conducted on Earth could become a commodity just as that done on Mars should.

All of this is impossible to quantify, but the history of free enterprise on Earth strongly suggests that some types of trade would develop. It would happen much quicker if humanity goes to Mars with the attitude that we are there to stay and eventually plan to pay our own way.

***

What sets our dreams for Mars apart from other human objectives is the sheer enormity of the endeavor. *We ultimately wish to see the Martian branch of humanity spread over the entire planet!* Remember, however, that the land surface of Mars is nearly equivalent to that of the Earth. It is necessary to take several steps back from that objective in order to identify something that we can relate to that will impose boundaries on our speculation.

For purposes of picking an intermediate objective to focus on, here are five postulated stages leading to the widespread settlement of Mars.

Exploration Stage: This might also be considered Settlement Stage 0 if a closely integrated exploration and settlement plan were in effect. Geologic exploration, the search for evidence of past or extant life, and scientific discovery are assumed to be the prime motivations for the mission, with settlement related objectives secondary. The following assumes that settlement needs, such as a plentiful water supply and appropriate local and nearby geological features, would be taken into account when the landing site was chosen. As per the earlier discussion, this is the best (and perhaps the only) way to assure that private capital and free enterprise will ever become a part of the future of Mars.

Settlement Stage 1: Permanent residents (Settlers) accompany explorers to Mars and are living at one or more exploration site(s) (now termed "Bases") in permanent housing. Each base supplies the basics required for livelihood and recreation while the Settlers install equipment and prepare for the arrival of more Settlers. Settlers are responsible for all base operations and expansion activities after exploration team departure. Members of the exploration team may elect to stay if sufficient settlement preparations make it practical and the Settlers agree to accept him or her.

Settlement Stage 2: Settlers assume complete responsibility for all functions at all bases/settlements; exploration teams from Earth phase out as Settlers begin negotiating exploration contracts with agents on Earth. The fledgling settlement(s) expand their facilities between exploration party visits, and new Settlers join them. Negotiated exploration contracts are fulfilled.

Settlement Stage 3: Settlements grow in number and population and expand trade relations with nations, companies, and people of Earth. Settlers assume control of immigration policy.

Settlement Stage 4: Collectively, the settlements achieve basic self-sufficiency. Trade with Earth continues to grow.

Let us expand on Stage 3 and use it as our baseline for discussion and analysis.

Assumptions for Stage 3 at Maturity

1. An economic system has evolved that is very similar to the one illustrated in Figure 2.

2. A common currency system is in place encompassing all settlements.

3. During the earlier phases of exploration and development, all supplies not available on Mars were

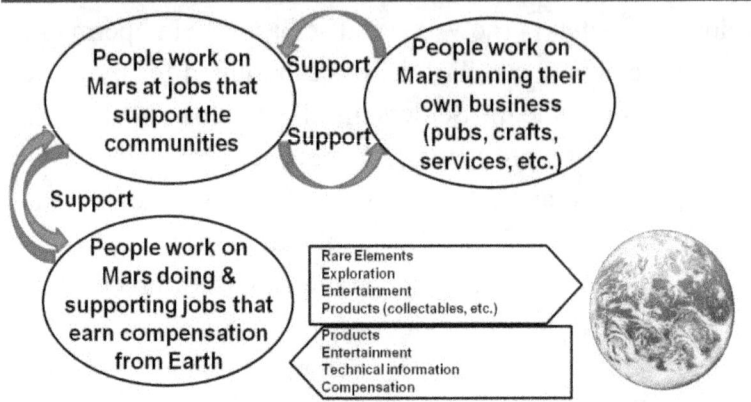

Figure 2. Simplified economic structure after Mars becomes self-sufficient

provided from Earth using cargo ships. As the settlements become more self-sufficient, imports from Earth begin to be dominated by personal items purchased by Settlers using currency electronically accumulated on Earth or Mars.

4. An entertainment industry is evolving based largely on Earth demand. Compensation contracts are negotiated and fulfilled.

5. Martian sporting events are becoming popular on Earth and compensation contracts are negotiated and fulfilled.

6. Mars Transfer Vehicles (MTVs) haul new people plus cargo to Mars orbit, and products are transported back to Earth orbit.

7. Any cargo that does not fit on the personnel MTVs, whether bound for Mars or Earth, will be shipped on separate MTV cargo vessels.

Stage 3 was chosen for further dissection because it represents the transition from exploration to settlement as the motivation for people going to Mars. The chosen juncture is also significant because the infrastructure for exchange with the Earth is all in place and complete. From that point forward, people will be making application to come to Mars largely based on one of the 16 motivations discussed earlier.

There is another significant factor with regards to selecting this stage of settlement development for closer examination. The Settlers would be in a position to partially or totally subsidize the cost of transportation for new immigrants if they choose to do so. They can do this by paying some portion of the new Settler's fare to get to Mars using the products shipped to Earth on the return flight. Electronic commerce (entertainment, exploration reports, etc.) could also be used to pay for a new Settler's travel.

This type of bootstrapping would become a very powerful tool for the settlements wishing to expand their population along strategic vectors. Settlers would be in charge of selecting immigration candidates. The process also goes a long way to dispelling the current notion that a ride to Mars would always be too expensive for anyone but the very wealthy.

Exports would expand as the population increases and those individuals, companies, or other entities that invested in the settlement of Mars would begin to see a return on their investment.

***

Any scenario that involves trade with Earth requires one or more commodities that have significant demand on Earth. Everything we know indicates that the *product* commodities discussed earlier are highly likely to exist on Mars. Unfortunately, no one can provide absolute guaranteed certainty until type, quality, and quantity of each product is subjected to assay. This applies to all types of precious stones and rare elements, especially those created by asteroid impacts. Unfortunately, some excavation will be required before the treasures buried in and around the craters are subjected to rating and evaluation. This will require equipment and time.

On the other hand, it is easy to speculate that sports and the arts would find substantial audiences on Earth. Not only is it logical, it also "feels" right after observing the trends on Earth in the entertainment business for the last 40 years. A 0.38g gravity field would guarantee a uniqueness of movement that cannot be duplicated on Earth without wires. In addition, adding to the lure of audiences would be the opportunity to witness firsthand the generation-by-generation changes in performances and performers as the species evolves.

One sure-fire revenue producer would be the market for exploration on Mars. This is an enterprise worthy of some discussion, including a bit of unbridled speculation, just for fun.

Mars has a land surface area of approximately 144,000,000 square kilometers, which just happens to be very close to that of Earth at 149,000,000 sq km. If 144 sq km (12 km by 12 km) were explored and documented every day, it would take one million days to cover the entire planet. That is 2,740 Earth years! Now, obviously, no one would be interested in thoroughly exploring every 12 km by 12 km plot of land on Mars, but even 10% of that area would require 274 years.

It is interesting to speculate on the potential income exploration might bring into the Martian coffers. If only one percent (1%) of Mars were explored under contract at $1 million per square kilometer it would generate $1.44X10^12 or $1,440,000,000,000.00 in revenue. That is over one thousand billion dollars. It is likely that many areas would be worth over $1million/sqkm to explore. Boom times on Mars!

From a larger perspective, it is logical that exploration would necessarily be focused by the intent of the sponsor. It would not be a matter of simply finding life or producing higher resolution maps for the USGS and Google. Some explorations would focus on searches for gemstones, diamonds, impactites, and similar rare stones. Others would seek deposits of rare elements and minerals of value, and there would always be the search for evidence of past and/or present life.

Eventually, the expanding industries on Mars would become exploration clients as well. Raw materials such as aluminum, copper, and iron would be needed on Mars, but it would never be economic to ship them to Earth as long as they are readily available here. This situation may change at some point in the future, as the raw materials on Earth become more difficult to get to and refine and Mars shipping gets cheaper.

Searches for organic materials and proof of past or current life forms would continue for a very long time, regardless of whether examples have previously been found. If one form is found then it follows that there must be others, and then the issue of older and newer specimens comes into play. This alone will fund explorations for untold decades.

You may be surprised to learn that agriculture could eventually become a major export, as Zubrin and others have pointed out. Clients could extend from the Moon and Earth orbit to the Asteroid belt between Mars and Jupiter. As anti-intuitive as it seems, with fully matured interplanetary transportation, it should be cheaper to transport a kilogram of goods from Mars to the Moon or Earth orbit than to ship it from Earth.

So what is our objective? Simply stated, it is one or more settlements that are self-sufficient sited in one or more suitable locations on Mars with healthy and thriving Settler families. Now we face a very tough question; is this objective achievable?

Health and Evolution Issues

As previously stated, humans who spend time in space experience physiological changes. The specifics and the significance of these changes vary between individuals and with exposure time. There is extensive discussion in space medicine and other literature concerning these phenomena. Physiological reaction to the new environment was not a surprise to many in the space medicine community, as it has long been known that almost all species react to changes in their environment. However, onset times and the level of effects have been generally regarded as unexpected, and there have been a few surprises. Let us look at some of the major issues that have been identified as they relate to getting to and living on Mars.

Gravity

Gravity affects everything that you do. Talk about something we take for granted! Gravity holds everything to the Earth including you, assuring that you do not fly away and blow up like a balloon in the vacuum of space. (Of course, gravity holds the atmosphere in place so if there were no gravity you would have already blown up, but that is just picky.) Gravity determines how much you weigh. Gravity decides how your clothes fit you. Gravity is partially responsible for how tall you are.

It also affects you in much more subtle ways. Gravity influences how your blood circulates. Your body has also developed a myriad of responses to and uses for gravity. For example, your brain knows which way is "down" because of gravity. Hips, knees, ankles, and the nesting of your internal organs all evolved as a response to Earth's gravity. The list literally goes on and on.

So, this thing we call gravity – what is it exactly? (Subtle drum roll, please.) No one, and I mean *no one*, has a clue. Some people would like you to think they have a clue. They are quick to jump into describing warped space-time and other vagaries of physics, including quantum spooky land, but when all is said and done, no one has a clue. If you wish to have your brain bent and perhaps partially destroyed, bring up your search engine and type in "what causes gravity" and hit enter. Good luck.

You are probably ready to ask, "What does this have to do with having babies on Mars?" Good question and the answer is "maybe everything" and "we do not know." The human species evolved in the gravity field of Earth, which we nerdy eggheads refer to as one gee (1g). Now, Sir Isaac Newton figured out a long time ago that everything that exists has gravity and these "gravity fields" attract each other, kind of like magnets, except that there are no opposite poles that attract and like poles that repel. There is nothing but attraction, and the more mass something has, the more it attracts everything.

How much you and the Earth are attracted to each other is called "weight." You and the Earth attract each other more when you increase your mass by having a large pizza every night for a month. Yep, more weight, but it is not that the Earth has changed noticeably, you have. Okay, now we get to the point.

Mars is considerably smaller than the Earth and has much less mass, so its gravity is less. Sixty-two percent less, in fact, which means that Mars gravity is 0.38 (38%) of Earth's.

Someone weighing 100 kilograms (220 pounds) on Earth will tip the scales at a mere 38 kg (84 lb) on Mars. You might say, "Hey that is a good thing!" The problem is, we do not know if it is a good thing, or a bad thing, or if it does not matter either way.

"Why are we concerned?" you might ask. There is a simple answer based on human responses when there is no *effective* gravity; bad things happen to everyone, and they begin to happen quickly. Before reviewing those disturbing observations, let me explain what I mean by *effective* gravity.

We are all familiar with images of astronauts and cosmonauts floating about on the space station, and it is not unusual for a commentator to refer to their gravity environment as "zero g" or "weightless." That is not correct, because the attraction of the Earth for anything 430 km (267 miles) above the surface is diminished by only a tiny fraction of the ground value. What is actually happening is that the space station is moving so fast that it literally "falls" around the Earth.

The people inside are in a state called "freefall" which is analogous to being inside of a freely plummeting elevator or on a crazy high roller coaster ride. As you might imagine, some people find this sensation of constant falling very rough on their mental and digestive systems, requiring a certain amount of time for adaptation. However, some people are not affected by nausea at all and almost everyone recovers fairly quickly. That is not a major concern.

To be a bit more picky, the state inside of anything in orbit is properly called "microgravity" because there are tiny forces (such as a very small amount of atmospheric drag) acting on the spacecraft. It is immaterial what the condition is called.

The point is that the occupants of the station are still well within the Earth's gravity field but everything behaves as if they were not, and therein lies the problem.

When a human body is suspended in microgravity, things that millions of years of evolution put into play have a new set of boundary conditions. Fluids that naturally flowed "downhill" suddenly have no reason to flow at all. Blood in the heart has no reason to exert a force on the lower portion of the heart, the muscles compensate by making the organ a bit rounder, and that becomes its new natural state.

The bones find themselves completely unloaded and some controlling mechanism decides that the skeleton is pretty much useless, resulting in an immediate response. The skeleton begins to decalcify. Muscles that normally fight gravity for 16 to 18 hours of their day have nothing to do; they get a pink slip and start retirement early. You get the picture by now.

A great deal has been written about the effects of microgravity on male and female astronauts. Even with a vigorous and specifically designed exercise regimen, everyone has had difficulty readapting to Earth's gravity after a few months in orbit. Again, the recovery time and specifics vary by individual. People spending five to nine months traveling to Mars may require an "artificial gravity" environment on the way in order to be viable upon reaching the Red Planet.

This can be provided since the effects of gravity can be duplicated by slinging the spacecraft in a circle at the end of a long arm or tether. The forces acting on anything inside of the vehicle will provide a simulation of gravity commonly known as artificial gravity. A later chapter will go into detail on artificial gravity and centrifugal effects.

A great deal of work has been done regarding rotating environments, and there are good reasons to expect that the overall effect will be to prevent the negative residuals experienced after extended exposure to microgravity. The problem is that implementing a rotating environment for the working and living quarters of a spaceship headed for Mars may impose increased complexity and cost. Those two words are very unpopular in management circles, and the reaction by the human space flight community around the world has largely been to throw money at understanding and mitigating the microgravity issues using pharma and mechanical solutions.

However, there is agreement by many in the scientific community to the effect that extended voyages lasting over a year must provide an AG environment for the astronauts. This *should* assure mitigation against all effects of microgravity, but no one can be 100% sure. Why not, you ask; because no one has ever tried it out.

As noted earlier, Mars has a surface gravity that is 0.38 or 38% of the average on Earth. Therein lays our interest and focus since the explorers may spend up to four years traveling to Mars, exploring, and then returning to Earth. In addition, settlement means that the emigrants from Earth will live the rest of their lives in the new "gravity well" and their offspring will spend all of their life there. As the following snippets allow, there are some indications that muscular degeneration can be slowed or halted by constant exposure to less than one Earth gravity. If so then perhaps other effects will be similarly attenuated.

Some of the Soviet scientists involved in studying reduced gravity effects (Shipov in particular) reported in 1981 that turtles and rats in Biosatellite Cosmos indicated that 0.3g was sufficient to prevent atrophic muscular alteration. Shipov also noted as an aside that Russian astronautics pioneer Konstantin Tsiolkovsky had estimated (method not reported) that 0.28g would be sufficient to maintain physiological integrity. Anecdotal, but fascinating nonetheless.

Based on these and similar results obtained experimentally, a group of former Soviet Union scientists declared in 1996 that the minimum effective artificial gravity level should be 0.3g. However, they recommended that 0.5g be used in designs to allow for some safety factor. There is no way of knowing at present if other lack of gravity effects observed on the ISS would be likewise mitigated at 0.3g or 0.5g.

Needless to say, those reports are not conclusive enough to risk the health and lives of astronauts and Settlers. Regardless of how the Settlers travel, those born on Mars would spend their entire lives in that 0.38g gravity well and it is highly probable there would be certain inescapable effects.

No Return to Earth

The Settlers would probably never be able to return to or visit Earth after some indeterminate time living in 0.38g. Exposure to home planet gravity would almost certainly overwhelm their skeletal, muscular, and cardiovascular systems. This is a potential issue for the exploration of Mars as well as its settlement. Exploration missions using current technology are estimated to take about 180 to 260 days to get from the Earth to Mars and about the same to return. There are quicker routes requiring faster rockets and more fuel, but they will not be available until someone develops the required passenger-rated propulsion systems. On top of that, of course, is the time they spend on Mars.

Even if the total mission duration, say for a flyby, can be reduced to 500 days, no one knows what it will take for an astronaut, one of the finer specimens of humanity, to return to normal activities after arriving back on Earth. Some mission scenarios place the total trip time (with and without landing on Mars) at over 1,000 days. Somewhere in this mish-mash of variables is a break point where the outbound and inbound ships will have to provide Earth-like pseudo-gravity in order for the explorers to return home to their hero's welcome.

So, what of the Settlers? The best guess I could find recently was that the challenge of transitioning from 0.38g to 1.0g will be dependent on the length of time an individual spends in low g and the exercise regimen followed during that time. We know from International Space Station experience that drugs and exercise cannot block all of the changes brought about by microgravity. Until we have data proving otherwise, it is impossible to know if the same will be true on Mars. Logic leans toward the 0.38g environment being less detrimental than microgravity, but we will not know for sure until the experiments are done on a rotating test and evaluation facility in Earth orbit.

For those who grew up on Earth, it seems highly likely that a shortened stay on the order of six months coupled with a strenuous and focused exercise program will help attenuate re-adaptation problems. However, there is no experience with extended stay at 0.38g and absolutely no reason to assume that a person living for years under those conditions will be able to return to Earth. An exoskeleton would help with Earth gravity effects on muscular changes but probably not with skeletal and cardiovascular issues.

Reduced Gravity and Physical Changes

We obviously have no insight into how the various bodily organs and systems will develop and function for anyone conceived, born, and raised in a 0.38g gravity field. It is easy to speculate that such offspring will be taller than their parents and that they will walk and run differently, but beyond that, all becomes guesswork. Keep in mind that this discussion only addresses gravity related questions at this point. Differences in atmospheric conditions, radiation, magnetic field, and diet will be touched on later.

There does seem to be general agreement in the literature that the cardiovascular system of the Martian children will undergo changes, possibly over multiple generations. There is no doubt that muscle and bone specific properties will evolve in response to the new environment of reduced stress and strain. Logically, we would also expect ligaments and other connective tissue to respond to humanity's environmental changes.

Such issues are daunting to say the least. If the scientists and physicians get the predicted effects wrong it will set the stage for one of the most horrific episodes in human history. ***This is an instance where evolution cannot be left to chance.***

Reduced Gravity and Neurological/Mental Function

This is a field where, again, uncertainty is rampant. The space neuroscientists know a great deal about the function of the central nervous system on Earth and have learned much about its function in zero-g in the last 53 years. They have some data on the 12 Apollo astronauts who visited the Moon, briefly experiencing 0.17g, and (obviously) have no data on human responses in 0.38g.

Humans are very adaptable to contrasts in environments as exhibited by their diverse locations on the Earth, from the Himalayas to the Australian bush country, and from the frozen tundra of the North to New York City. Unfortunately, for our purposes, they all live in a 1.0g gravity field. (Okay, so that is not exactly accurate since gravity does vary with changes in latitude and altitude by as much as 0.2%. Give me a break!) Adaptation to microgravity and 0.17g environments has been analyzed (to varying levels) and it seems certain that Settlers on Mars would likewise adapt, but will there be a toll in function, efficiency, or longevity? No one knows.

It seems intuitive that the human response in 0.38g will be somewhere between that in 1.0g and microgravity, that is, for the Settlers born on Earth, but what about those conceived, born, and raised on Mars in 0.38g? As the scientists often say, more research is required. One thing we can be certain of; multiple attributes of these Martian humans are going to be different compared to their Earther cousins due to the differences in the gravity fields. We just do not know exactly what and how much.

Radiation

The subject of radiation dosing for astronauts involved in any type of trip beyond low Earth orbit (LEO) has become charged with emotion and muddled by misunderstanding. I will not attempt to sort it out here or even discuss the issues. My rationale for dismissing the issue is as follows. *People are not going to settle down, live the rest of their days, and work their butts off where they cannot bear children and raise families.* There will never be any long-term expanding settlements on Mars or anywhere else unless radiation levels are achieved that are acceptable to childbearing Settlers.

If that level of safety is not achieved, there will never be more than exploration missions to Mars, possibly along the lines of Antarctic expeditions. This means that one of our fundamental assumptions must be that *safe radiation levels will be guaranteed to Settlers*. Fortunately, this is not difficult to achieve on Mars. Reaching significantly reduced levels in a spacecraft in route to Mars is more difficult but it is achievable. However, an *exact duplication across the spectrum* of what Settlers would be exposed to if they remained on Earth is out of the question.

All we can infer from this is that from sperm and egg to full-grown adult, a human born on Mars will be exposed to a different spectrum-intensity distribution of radiation than his or her opposite on Earth. There may be subtle effects at first with more pronounced changes manifested in later generations, or there may be none. Right now, we do not know. This will remain an area of uncertainty until multiple generations have propagated on Mars.

Atmospheric Pressure and Composition

It is desirable to reduce the pressure in the habitats to about one-half of Earth sea level normal since this has a dramatic effect on the weight of the structure and reduces the quantities of required gases. This is one area where there is substantial knowledge about the response of the human body to changes in atmospheric pressure and composition. What we do not have is a database on how those changes in pressure and composition affect child development and growth, especially in the presence of the 0.38g gravity field. This is an uncertainty that could be evaluated in the simulated Mars environment discussed later.

Lack of a Significant Magnetic Field

Mars has a very weak magnetic field located only in its southern hemisphere, whereas the Earth has a field several thousand times stronger that encompasses the entire planet. Among other useful features, our neighborhood magnetic field helps protect us from cosmic and solar radiation by deflecting much of it around the planet. There appears to be no solid scientific evidence supporting any direct interaction of the Earth's magnetic field with human physiology or psychology. However, it is well established that some animals navigate by sensing the magnetic field. It is unknown whether developing and growing without a magnetic field will affect children.

Aerosols

Apollo astronauts fought an enemy on the Moon that was more vicious than anticipated; dust. Lunar dust was very fine but abrasive, and it stuck to anything and got into everything. This was partially due to the jagged nature of the surface of individual particles, which is in contrast to the dust on Mars. Martian dust is very fine but generally has a smooth surface due to eons of being blown about. This makes it much less abrasive but not necessarily less "sticky."

In addition, some of the dust on Mars is chemically active and toxic. Exposure will obviously have to be controlled, but it may not be physically possible to eliminate all exposure in all circumstances. Safeguards in the form of counter-medication or similar measures will probably be required, at least in the early days.

Regardless of how dust and its effects are controlled, growing children on Mars are going to be exposed to a unique aerosol environment whose effects are not completely understood. In addition, the 0.38g environment affects the behavior of dust in the rarified atmosphere of Mars. Unfortunately, this is one area where experiments conducted in artificial gravity generated by a rotating environment may not yield much data of use. This will be obvious later when Coriolis Effects are explained.

Diet

Although not an area of major concern, the foods consumed by the children of Settlers will influence their development and possibly certain evolutionary trends. There may be analogs to study on Earth, and this would provide a fascinating area of research.

Other Health Issues

This is a catchall. Degradation of visual acuity has been permanent in several astronauts. Increased cranial pressure is the current theory of cause and more research is in progress on the ISS. A number of uncertainties exist relative to cardiovascular function and bone calcium loss as a function of gravity and activity. The list of concerns is long and the list of uncertainties is even longer. One of the more troubling concerns the population of microbes and tiny beasties that we live with everyday and never know it. How will these new environments affect them and in return, what effect will that have on humans? No one knows.

Will we ever have all of the answers, resulting in an ironclad guarantee to every explorer and Settler candidate that they would be in no more danger than sitting at home on Earth? Of course not. Traveling great distances in space and spending long periods in an alien environment will always carry risk. What we need to be sure of is that the risks are understood and acceptable, especially in the context of children and future generations.

Evolution

The first parents on Mars must to be ready to accept that their children will be different from their cousins on Earth. Each generation must do the same. A greater challenge may be gaining acceptance on Earth of this inevitability. The very fact that it is going to happen may upset a significant percentage of the human race and create resistance to the idea of settling Mars. Of course, if it turns out to be impractical or impossible to settle Mars then it will be a moot point.

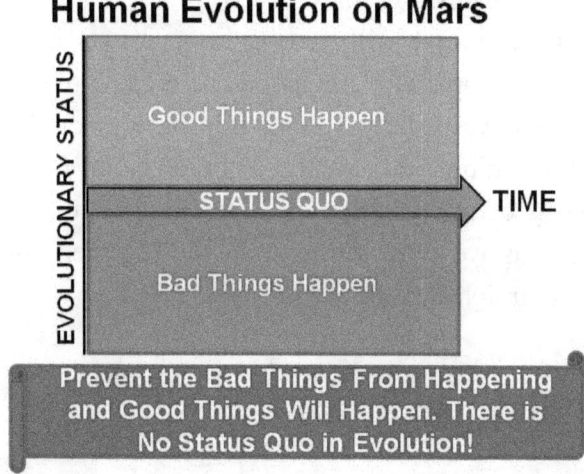

Figure 3. Controlling evolution may become the greatest challenge to settling Mars.

Given the distance to Mars, it is doubtful that the Martian branch of the tree of human evolution would be perceived as a threat by a substantial percentage of the world's population. More troublesome might be the reactions of religious leaders.

The most important challenge to Mars settlement advocates is assuring that the evolutionary changes occurring on Mars will not be degenerative. There is no status quo where evolution is concerned. When the environmental changes are as dramatic as we are discussing here, either good things will happen or bad things will happen, as illustrated on Figure 3. It is essential that we determine if the potential for bad can be negated. Finding out the answer by trial and error on Mars is absolutely *not* acceptable. If degenerative evolution can be prevented, then we would significantly increase the probability of positive enhancements happening to our Martian cousins.

The one area that we will have no control over is gravity on Mars. In space, artificial gravity can be created to simulate that of Earth and Mars, but on Mars, it is simply not practical to house everything and everyone in a giant centrifuge creating one g of artificial gravity. Herein lies our greatest unknown, but as we shall see, getting the required answers could be very straightforward.

Why Am I Telling You This Now?

One thing to remember is that ANY commitment to "explore" Mars is not a commitment to settle Mars and research that supports exploration does not necessarily support establishing settlements. If money is going to be spent to conduct human exploration, it would seem wise to make the technology, hardware, and results applicable to a permanent presence. However, that is only a part of the answer. More importantly, _there are areas of research that are critical to whether Mars is ever going to be settled that are not of particular interest for exploration planners_.

Pregnancy, birth, and the growth of children are prime examples. The research necessary to determine from the Settler's point of view what is achievable and what is not will be expensive and controversial, and it will take a very long time to come to fruition. We need to begin the dialog as soon as possible to assure that issues relevant to settlement are seriously considered in the design and operation of any space-based experimentation and simulation facilities. This is especially true for those built to address gravity related uncertainties. NASA seems to have no interest in pursuing these topics and actually seems to actively avoid them.

Interest in exploring space, Mars in particular, has recently been on an upswing for a variety of reasons. Near the top of the list of motivations molding public attention is most certainly the Mars Science Lab, better known as the rover named Curiosity. The suspenseful landing, the drama free checkout, and the brilliant execution thus far of its intended mission have done wonders to attract positive attention. The recent measurement of a transient methane gas concentration by Curiosity's sensors has excited the media. Determining whether the methane is the product of organic material processes, geologic phenomenon, or rover flatulence may have occurred by the time you read this. If the science community can all agree that the source is organic, Curiosity will probably be named Person of the Year! On the other hand, if (as some suggest) the methane turns out to be a by-product of the rover itself, embarrassment may again replace big smiles.

Another stimulant for the increased press coverage has been publicity surrounding the Mars One project described earlier. Controversial, yes, but be assured that these folks are serious. No entrenched expert (including yours truly) will give them a very high probability of success, but I do try to help them in my own modest way, and I wish them luck. Tens-of-thousands of people around the world support their mission, and that number is probably conservative. There has certainly been no lack of credible and apparently sane volunteers to make the one-way trip. The (non-technical) problem is that they need American dollars by the billions, not tens-of-thousands, or hundreds-of-thousands.

Yes, that was billions with a b. They are currently seeking funding from multiple sources, including investors and a "reality" TV show about the astronauts (as they call them) during their training and subsequent trip to Mars. It is also correct to call them Settlers and Settler candidates because they meet the basics of our definition. The move is one-way and for the duration of their lives with the intent to expand the first base into a community.

Fantasy, you might say, but in fact, Mars One is the only game in town as I write this. (More about Elon Musk and

SpaceX later.) Mars One has declared their intent to settle Mars; they have raised money and issued contracts for hardware design; they have taken applications and are screening candidates to be trained as Settlers; and they continue to pursue further funding as they narrow down the selected would-be Martians.

None of the manned Mars mission touting countries (USA, Russia, and China) has political leadership that is committed to funding such a journey. Nor, at the time of writing, is there an active program office in place that is working in a focused fashion and spending money to achieve program milestones that will result in a human mission to land on Mars. In addition, none of these three have professed to studying (much less advocating) a one-way trip for anyone.

However, the big three all include a landing on Mars in their "long range plans" which encompass the 2030s and 2040s. I have lived through fifty years of government "long range plans" for space that never went anywhere, and so far, I see nothing in this century to change that trend! NASA, the Russians, and the Chinese are talking a good game as I write this, but I will believe it when I see real commitments made.

Eugene Lally expressed the sentiments of our generation well in the Sep/Oct 2012 Issue of "Space Times" published by The American Astronautical Society. *"We spent the past forty years fooling ourselves that we were advancing toward manned Mars missions by spending our time at related conventions, conferences, organizing international partnerships, in committees, in development collaboration, forming agencies, charting the way forward, forming long term goals and cooperative frameworks, implementing incremental steps using dedicated programs, planning to* *adapt to changes in international partnerships, etc. **All this posturing and time spent with no focus or leadership for decision making led to no results.*** Emphasis added by this author.

Unfortunately, this is still going on. In spite of NASA's use of the Mars word whenever the opportunity presents itself, the United States does not have a humans to Mars program.

Another reason for the recent attention on Mars has been Elon Musk, the flamboyant yet introspective founder of PayPal, Tesla, and SpaceX. Musk has repeatedly expressed his support for populating Mars, even to the point of stating that he founded SpaceX as a company to reduce the cost of transportation to the fourth planet. Musk and his team are spending a portion of his fortune and the profits from his company to make a trip to Earth orbit affordable, so we know he is serious.

SpaceX has begun to work on the interplanetary transport concepts and Musk has said that he will be unveiling his plans in late 2015. By the time you read this, SpaceX may be the second game in town. Elon Musk is my new space achievement hero. I hope to get an autographed picture one day to put alongside the one I have from Dr. von Braun.

There are myriad other efforts under way by companies, institutes, organizations, and individuals. Typing "settling the planet mars" into your search engine will bring you links to all or most of them. Be prepared to settle in for a long session of search, download, and read. Under the circumstances, I feel comfortable with suggesting that NASA, Mars One, and SpaceX will be the three primary players in the headlines over the next five years. However, many of us would welcome a wild card or two in the mix.

Why is NASA in the company of Mars One and SpaceX in my short list you might ask, after noting above that they have no Mars program. I list them as an actor because they will be funding much of the key technology work. Various NASA planning charts that have been made public show some current or planned technology programs that relate to Mars exploration. Keep in mind that even a robust exploration technology program (which does not exist at present) will not address all of the settlement issues. As was stated previously, an excellent example concerns the issues of progeneration, which prompted this book. The following issues must also be addressed from a more mundane but no less important perspective.

Settlement must build directly on exploration to avoid being at a great disadvantage. However, there will be a tendency for the exploration planners to evaluate mission locations based strictly on scientific merit unless someone speaks up.

The following are examples of the types of factors that should be considered from the settlement perspective in addition to pure scientific criteria.

Water: It is well known that certain zones on Mars have higher water content in the regolith permafrost than other regions. In addition, there is extensive penetrating radar data indicating that entire glaciers of "dirty water" are lying hidden below layers of dust, dirt, and rubble. The latter has not been confirmed through in-situ inspection, so we have no direct proof of the existence of these glaciers or any idea what the chemical makeup of the ice might be, but there appears to be a lot of it.

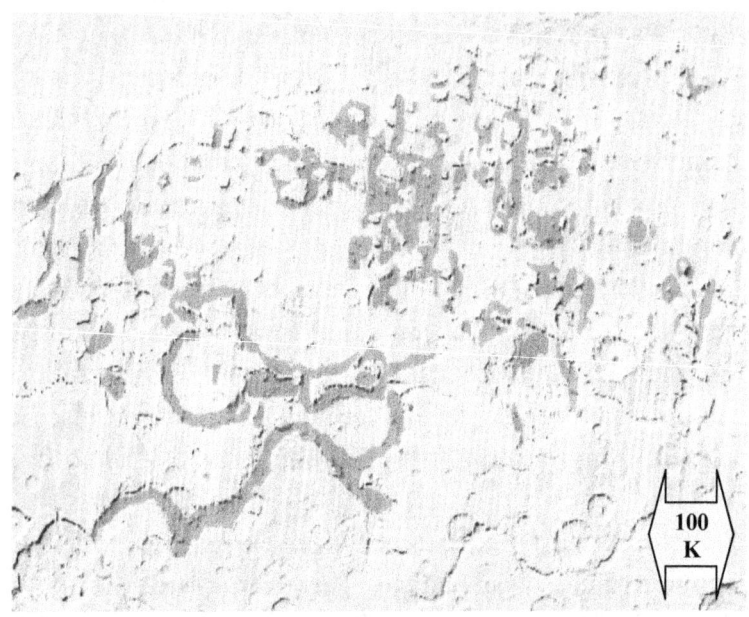

Figure 4. Dark areas designate locations of glaciers, some believed to be one km thick

The glaciers of Deuteronilus Mensae are enormous. Using the published interpretation of the radar data illustrated in Figure 4, I calculate that there is on the order of 1,000 trillion cubic meters of "dirty water" available there. That is approximately 260 X 10^15 gallons of water, enough to last one million people indefinitely with recycling. (Once the number gets over 100,000 years I just treat it as "indefinitely." Actually, it would still be "indefinitely" without recycling!)

That is only an example. Covered over glaciers are indicated by radar and geology in numerous locations so selecting an area satisfactory to exploration and settlement should not be difficult. Many Mars settlement concepts rely on water being harvested from ice-laden permafrost, which is certainly feasible and practical for small settlements. However, that approach is self-limiting and it will not sustain substantial population and industrial growth.

Soil: Examples of interesting locations near probable glaciers abound, and other factors would need to be entered into the basing location selection. The Settlers would be particularly interested in ready availability of soil suitable for modification and integration into intensive farming. Precursor stationary or mobile robotic probes may be required to test soils in an area to assure that they are compatible with the requirements for agricultural applications before the exploration mission location is finalized.

The utility of regolith for construction and/or radiation absorption purposes will also need to be assessed, along with the subsurface strata capability to carry weight and provide anchorage. These are only minor examples. Proper engineering work would need to be done to identify the parameters requiring investigation during the first exploration landings.

Local mineral sources: It has been shown on Earth that common and rare minerals are both found in and near meteorite and asteroid impact craters. These types of resources would factor strongly into settlement site selection and ultimate fiscal viability. Exploration planning should include availability of craters of interest and some means to assess potential mineral sources.

Solar insolation: Having certain levels of solar insolation throughout the Martian year will be desirable and should be taken into account in the site selection process.

Local terrain suitable for expansion: Expansion of a permanent settlement should be planned for from the very beginning. Certain criteria for surface condition and irregularities as well as proximity to other expansion areas would be among the considerations.

***

This is by no means an exhaustive list, just a few examples to make the point that synergism between exploration and settlement planning can have far-reaching implications. This will be especially true if the new capabilities in construction afforded by the environment on Mars are taken into account.

So, why spend time contemplating Mars settlement issues when their implications seem so far in the future? Someone must work to *insure that the research and pioneering work in the exploration phase does not leave the questions related to family building on Mars as open issues, thus turning a long shot into an impossible dream for another generation.* This is especially true in the 0.38g research arena. That work must include a long-term study of pregnancy, birth, and growth of mammals, especially primates.

The 400 kg Gorilla

The cost of transportation to and from Mars is often cited as the ultimate reason that settling Mars and establishing trade could never be feasible. "Trade" in that context means the swapping of physical goods from Mars for physical goods from Earth and other forms of commerce. As was discussed previously, a substantial portion of "trade" items coming from Mars would be in the form of electronic media, i.e. exploration reports, sports broadcasts, entertainment, etc. However, ignoring the desire for trading products with Earth would be like trying to ignore the proverbial 200 kg gorilla in the room; it cannot be done. One major reason would be the needs and desires of Settlers.

In *The Earth-Mars Chronicles,* I had the latitude to form an international Alliance that kept both Settlers and a supply train going to Mars on a regular basis. This made it possible for the settlements to expand and mature, and eventually become self-sufficient. Some similar scenario would be required to build permanent and growing settlements later this century. It is generally accepted that self-sufficiency would only be reached after substantial growth and diversification of the population accompanied by a similar expansion of manufacturing capability.

I have seen "guesstimates" that stretch from a population of 50,000 to about one million Settlers to achieve complete self-sufficiency. There are many assumptions that factor into making such prognostications and a great deal more study needs to be done, keeping in mind that the result will be highly technology level assumption dependent. Suffice it to say that a large population supported by expansive manufacturing and agriculture will be required to achieve self-sufficiency.

Given a requirement to have such a population, sending a few Settlers to Mars every 26 months or so does not measure up when the objective is to establish a viable, permanent, and self-sufficient branch of humanity. By a "few Settlers" I mean anything under 1,000 at every two year opportunity. I realize that fewer Settlers may be sent in the earlier stages when facilities are being built and expanded for those coming later, but remember that we are looking specifically at Stage 3 where settlement populations begin rigorous expansion.

That is a mind-bending number at a time when we cannot seem to reach a consensus on how to send just four people to the Red Planet! Even at 1000 Settlers per two-year opportunity it will still require approximately 80 years (exact time would be dependent on birth rate) to achieve a population of 50,000.

Ideally, Settlers and some cargo will make the trip from low Earth orbit (LEO) to low Mars orbit (LMO) in a ship that can return to LEO loaded with cargo. It would also be ideal if the value of the goods shipped back to Earth were sufficient to cover the cost of operating the Mars Transfer Vehicle (MTV) for both legs of the trip, thus paying the way for that batch of Settlers.

One thing becoming obvious is that an operational space infrastructure will be required to enable any "routine" interactions between residents of Earth and Mars. The MTV will be far too large to be built and launched from Earth. It must be built and assembled, or at least assembled, in orbit. Additional infrastructure will then be required to outfit and test the large interplanetary vessel, followed by fueling and additional testing. Similarly, cargo modules have to be moved about and attached to the MTV structure in LMO. In LEO, tugs must remove those cargo modules and transfer them to some type of landing system. Every tug must be ultra-reliable, capable of being refueled, and reusable for an indeterminate period with regular maintenance and updating. Most (perhaps all) of these functions could be executed by remotely controlled robot tugs.

Settlers boarding an MTV headed for Mars will have to have a ride up from Earth's surface, and conversely, they will require a ride down to Mars after arrival. Sending all of the required propellants up from Earth would be prohibitively expensive. This strongly supports the notion that asteroid and/or lunar miners (I dub them Roids and Lunies in my books) would have mature operations in effect when Settler MTV flights begin. This assumes, of course, that water from asteroid and/or lunar mining would be much less expensive than the equivalent hauled up from Earth.

At this point we have established a divergence from the current popular notion that "exploring Mars" should be done from Earth by people riding huge rockets to orbit and there meeting up with the payloads of other huge rockets. Everything would then be docked together and blast off to Mars, where equipment previously launched by huge rockets would be presiding, and automatically taking care of business. All that would be returning from Mars are the astronauts and a few kilograms of rocks. That mission architecture is Apollo on steroids, and that is not a particularly robust scenario for anyone dreaming of establishing a permanent settlement. Buzz Aldrin and Leonard David expressed the sentiment well in *Mission to Mars: My Vision for Space Exploration* (2013) National Geographic Society.

"All this is preface to a major judgment— one that I feel NASA planners are dodging. There is no reason to make a humans-to-Mars program look like an Apollo moon project."

On the positive side, there are companies formed specifically for the purpose of mining the near Earth asteroids (NEAs) and the Moon. Tugs, transports back and forth to Mars, Lunar shuttles, satellites, and miscellaneous Cis-Lunar traffic operators would all be in the market for propellants and other commodities the lunies and roids would provide. This is another critical piece of our Martian settlement jigsaw puzzle.

At this point, we current dreamers join forces with those of past generations. We want a permanent, operational, robust, and enduring infrastructure in space, one that could potentially serve the needs of a broad spectrum of space initiatives, including the exploration and settlement of Mars. On a much grander scale, realizing this dream will act as a powerful enabler to support human expansion into the solar system. *That is, unless our research dictates that children must always be born and raised on Earth.*

We, as a nation and as a species, need to step off the planet to stay. To do so requires that we do things in space beyond low Earth orbit that are more permanent and meaningful than lassoing a space rock and playing with it.

Effects of Reduced Gravity – The 400kg Gorilla in the Room

As bizarre as it seems in retrospect, a series of people in NASA management have managed to ignore this 400 kg gorilla for decades. In 1975, as part of the required work for a post-graduate course in Bioengineering, I did a literature survey on human adaptation to artificial gravity and to rotating environments. Most of the work had been done in Russia. In 1976 and 1977 my research paper was expanded to discuss the effects of rotation on people living in O'Neill space colonies. Gerry and I wrote and published a paper on that work in 1977.

(Gerard K. O'Neill and Gerald W. Driggers. "Observable Effects In and Human Adaptation To Rotating Environments." *Space-Based Manufacturing From Nonterrestrial Materials*, pages 173-176. Edited by Gerard K. O'Neill and Brian O'Leary. American Institute of Aeronautics and Astronautics, 1977. Volume 57, Progress in Astronautics and Aeronautics: technical papers derived from the 1976 Summer Study at NASA Ames Research Center.)

When I began research for *The Earth-Mars Chronicles* series a few years ago, I was somewhat eager to look into what progress had been made on "artificial gravity" in the intervening 38 years. The results astounded and appalled me. Little progress had been made and some of the research facilities had been shut down. Very little meaningful additional work had been done on the use of artificial gravity or the effects of rotating environments! You will see later that this trend continues.

***

As discussed at length in an earlier chapter, considerable work has been done on microgravity's negative effects on different systems of the human body. The following is a recap that leaves out the nuances and minor effects, which are plentiful.

1. Intracranial pressure with vision alteration
2. Fluid retention and ramifications

3. Back pain during adaptation

4. Kidney stone formation

5. Skeletal degradation (loss of calcium and other minerals)

6. Reduced muscle mass, strength, and endurance

7. Reduced air volume capacity

8. Periods of dizziness and disorientation, especially during microgravity adjustment period

9. Cardiac rhythm disruption

10. Difficulties associated with standing and walking after returning to Earth gravity

11. Reduction in immune system response

The astronauts and cosmonauts largely recover from these effects after a few weeks or months back in surface gravity with the exception of vision problems. Some have required glasses to correct their vision anomalies.

Research related to hypo-gravity (hypo-G, defined as anything less than Earth gravity but greater than no gravity) has been sporadic around the world and very nearly non-existent in the United States. The following rather lengthy quote is from the report of the *2014 International Workshop on Research and Operational Considerations for Artificial Gravity Countermeasures* held at NASA Ames Research Center on 19-20 February 2014. The following is excerpted from the Introduction to the final report (NASA/TM-2014-217394). All underlining has been added by this author.

***

"In the executive summary of the "Proceedings & Recommendations" of the AG Workshop 15 years ago in 1999 in League City, Texas, we stated that *"More than 30 years of sporadic activity in AG research has not elucidated*

the fundamental operating parameters for an AG countermeasure. For this reason, we do not advise NASA to discontinue support of countermeasures under development. Instead, we recommend that NASA appropriate the resources—primarily deploying and funding a peer-review research program—necessary to initiate AG parametric studies on the ground and in flight. Such rudimentary studies would serve as a basis for exploring an AG countermeasure and must precede prescriptions for the application of AG during long-duration space flight." Finally, we concluded *"our final recommendation is that NASA establishes a standing AG working group. The group would meet annually for the purpose of continuing and advancing our progress."*

The Introduction continues. Keep in mind that this was written by a team of experts in 2014.

"The only difference from then to today is that 15 more years have elapsed. The above statements are as valid today as they were back then, except the opening statement could be "More than 45 years of sporadic activity in AG research has not ..." Because NASA's vision for space exploration includes some nine design reference missions to send humans into deep space for long-duration (years) periods, the selection of the final health protecting countermeasure suites should include considerations for AG. The unique feature of AG is that it protects not just one but all of the physiological systems against low gravitational loads (hypo-G). For the time being, protective countermeasures are being developed to target specific physiological systems, which may be protective for one system, but with less or no protection for other systems. In addition, gender differences and individual differences exist in the response to various countermeasure interventions, which further complicates development of efficient countermeasure suites. <u>AG has none of these drawbacks, because all humans have throughout evolution adapted to the same 1-G level.</u>

"One possible reason why NASA has not seriously implemented AG as a health-protecting countermeasure during spaceflight is that the development of AG in space faring vehicles is perceived as being too expensive and complicated, from an engineering standpoint. Multiple studies, however, have shown that this argument may not be valid. Also, in the intervening years of research, NASA has gained insight into the efficiencies of our currently used countermeasures—in particular from utilization and research on the previous Mir space station and now on the ISS—so that a trade-off of these against implementing AG can be implemented on a more mature basis. This was the reason for reconvening this AG Workshop at Ames Research Center on February 19-20, 2014."

***

This makes for quite a strong opening for the report and the details in the document back it up. Conclusions and recommendations follow. The underlining has been added by this author.

"The main conclusion from the Workshop is that AG during long-duration space missions are feasible from an engineering perspective, and that three types of scenarios should be considered: 1) centrifugation inside a space vehicle; 2) spinning part of a vehicle; or 3) spinning the whole vehicle. <u>Research should be initiated as soon as</u>

possible to establish the life science AG requirements such as G-levels, durations, and centrifuge size, and in regard to whole-vehicle spinning the minimum G-level (threshold). In addition, the extent to which current countermeasures need to be combined with AG must be determined."

***

"Countermeasures" is the term applied to attempts to prevent a variety of these degradation issues. The word is plural; one thing most in the space medicine community seem to agree on is that there is no single countermeasure that solves every problem, with the possible exception of "artificial gravity." This point is given further substantiation through the following extract from a paper titled "Another Go-Around: Revisiting the Case for Space-Based Centrifuges" published in *Gravitational and Space Biology*, pg. 66, Vol. 25 September 25, 2011. (NRC is National Research Council; IAA is International Academy of Astronautics)

***

"The need for space-based centrifuges for both research applications and astronaut countermeasures has been articulated for decades. Key reviews and reports from NASA and the space life sciences community have long identified artificial gravity (AG) facilities as a top priority for the gravitational biology and aeromedical communities:

"There was unanimity of opinion that any major adverse effects of spaceflight could probably be prevented by creating an artificial gravitational field within the

79

spacecraft... The need for a centrifuge on future flights is of the highest priority
(NRC, 1979).

Variable Force Centrifuge...is the single most important facility in any life sciences program... [and] should increase the scientific return from space experiments by orders of magnitude...A VFC is an essential instrument for the future of space biology and medicine
(NRC, 1987).

Whether used in the near-term to facilitate human missions to Mars, or put off until developing missions to destinations farther away, artificial gravity will eventually be required to protect humans exploring space.
(IAA, 2009)."

Given the apparent preponderance of expert opinion, one might ask why so little work has been done on artificial gravity in the past 45 years.

There are multiple answers, but the primary reason has been priorities based on funding. Officially, the answer has been that the environment most suitable for doing interesting science in space is microgravity, so AG was not considered a requirement for the International Space Station (ISS). Providing a working environment of microgravity plus an off-time and sleeping area of artificial gravity would have seriously impacted the cost and schedule of the ISS. However, one of the objectives of the ISS became evaluation of countermeasures for human body microgravity degradation, and that continues to this day. A small centrifuge has been added to support some very limited experiments with insects and rodents, but serious AG research does not appear to be on the horizon, based on NASA's FY2015 budget breakdown and the comments of Dr. Julie A. Robinson on The Space Show, 31 July 2015. Dr. Robinson is Chief Scientist for the International Space Station. To summarize her comments, *artificial gravity is a poor solution left over from the von Braun era and it will not be pursued in the foreseeable future*.

Speaking of budgets and technology programs, you might ask why there are so few Mars expedition technology initiatives in the current and planned budgets. Good question, given all of the "Journey to Mars" hype.

Technology programs generally find their way onto planning charts when the advocate can link his or her project to one or more currently funded programs, some nifty new buzzwords, and/or popular future initiatives. Promoting support from the Congressmen and Senators whose districts and states will benefit from the funding of that technology effort is strictly forbidden, but it is always a plus.

However, getting a technology effort listed on a long-range planning chart holds no sway unless it shows up as an on-going program or new start within a current or near-term budget planning cycle. Then the fighting and dickering has only just begun. Every technology development or demonstration effort that does not have an established and funded program office fighting for it is automatically at a disadvantage. To my knowledge, there is no Humans to Mars Program Office fighting for coordinated and focused technology programs as of this writing.

This approach to selection and execution of technology programs leads to continuity and integration issues, especially for an initiative that is supported by vagaries. An excellent example is the lip service being given to a human mission to Mars "in the mid-30s." If President Kennedy had committed the United States and NASA to landing on the Moon and returning safely "sometime in the mid-70s," Apollo 11 would never have happened, even to this day.

Lip service is not a substitute for realistic commitments, especially in the arena of supporting technology development and demonstration. A trip to Mars, even one way, requires a tremendous effort in requirements definition, component and subsystem design, scheduling, integration, system level testing, and thousands of other activities. Those are the kinds of things that program offices make sure are done in an integrated fashion. Otherwise, all you usually get is a mish-mash of people's pet projects. To my knowledge, an adequately funded program office for exploring and settling Mars does not exist in the United States or elsewhere.

Budget constraints also explain why vessels voyaging to Mars in NASA architectures have never incorporated provisions for artificial gravity (AG). Conventional wisdom has it that incorporating a rotational capability in the Mars transit vehicle design imparts complexity, meaning more technology development and demonstration, which in turn means more money. To test this common assumption, one study has been executed to quantify the impact that incorporating AG would have on a deep space interplanetary craft. The report is "Preliminary Assessment of Artificial Gravity Impacts to Deep-Space Vehicle Design." It is available on-line. Author of the report is B. Kent Joosten, and the Document No. is EX-02-50. The study was published in 2005.

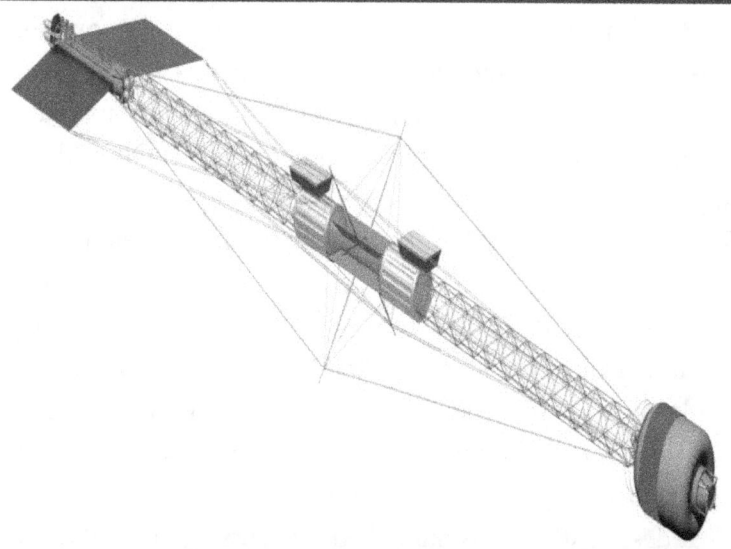

Figure 5. NASA JSC design for an Artificial Gravity Mars Transit Crew Ship

Under this mission concept, the crew ship, shown in Figure 5, would be the only one with AG. Pseudo-gravity equivalent to Earth (one g) would be generated by rotating the entire ship at four rpm. The atomic fission power plants were located at one end of the truss structure and the crew habitat at the other. Electric Ion propulsion systems and propellant were located in the middle on either side of the axis of rotation.

Everything else, including the lander, would make the journey to Mars in standard configuration non-rotating ships. The study team found that designing for 1g by incorporating rotation imposed some design compromises but did not have a substantial impact on the estimated mass. As you might imagine, having artificial gravity simplified a number of personnel items such as food services and toilets, except that many functions would need to operate in both microgravity and AG. On the other hand, the astronauts point out that micro-g simplifies a number of things, such as almost never having need for a "floor" or a place to sit. Lack of gravity also frees up wall space and simplifies moving about. Oh, and it is "just plain fun."

The biggest impact to past operational modes of this particular design, and all other designs incorporating AG, is the necessity for doing assembly in orbit. It is worth noting that it is not only the assembly but also all of the integrated testing that must then be done in space. However, as stated before, humanities expansion into our solar system demands that new operating modes must be mastered.

That design study was published 10 years ago in 2005. The AG ship was designed for one g because there was no data available to support selection of any other cruising condition. Note that nothing has been done in the intervening years or is planned regarding determining the effects on human physiology of long-term exposure to pseudo-gravity in a rotating environment. Any similar study done today, 10 years later, would be forced by the lack of data to make the same assumption.

We intuitively feel that 0.38g has to be better than microgravity, but will it be good enough to negate all of the negative health effects? Figure 6 illustrates the current dearth of knowledge. The only way to fill that vacuum is by using an artificial gravity research station in orbit about the Earth where multi-month and multi-year exposure studies can be conducted.

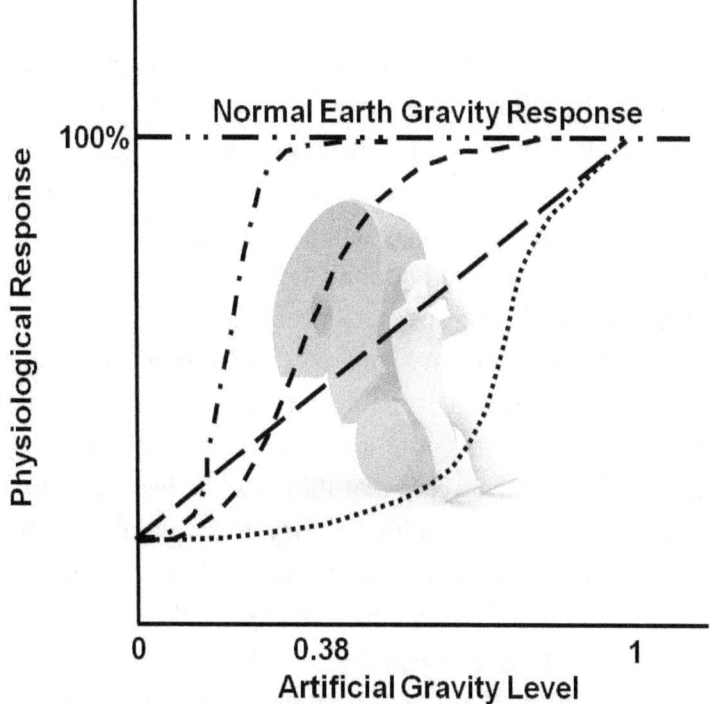

Figure 6. A disturbing lack of data and experience

A Closer Look at Imitating Gravity

In 1907, Einstein formulated a new concept called the Principle of Equivalence as part of making his General Theory of Relativity work in a gravitational field. One result of his revelation was the observation that a body or system of bodies being accelerated by some force respond precisely as they would in a gravitational field. In other words, the response of the human body to acceleration in a straight line at 9.81 meters per second squared (9.81 m/sec^2) in a rocket is precisely equivalent to being on the surface of the Earth. This may seem a bit trite to us now, but Einstein would comment in his later years that this was the happiest moment of his life because it allowed him to explain the Special Theory of Relativity in a gravitational field. To each his own.

The Principle of Equivalence is important to us because in space it allows us to substitute acceleration for gravity. Of course, if we tried to do that by running the space ship motors all of the time we would exhaust our fuel (catch the pun?) rather quickly. Instead, we need a simple method to induce an acceleration that will require a minimal expenditure of energy. Thus, we arrive at the "can on the end of a string" analogy.

A can half-full of water and attached to a string will not spill a drop if you swing it around in a circle quickly enough. At any given moment the water and the can both wish to go in a straight line and that is indeed what will happen if you cut the string suddenly. There is a force constantly being applied to the can and subsequently to the water due to the constraint of the string, and according to Sir Newton, that means that the can and its contents are constantly being accelerated along the axis of the string. If you are holding it as it swings, you will feel the increase in force as you increase the revolutions per minute.

That is how we can imitate gravity, by creating an "equivalent" acceleration through rotation and constraint. This was realized by the rocket and space travel pioneers of the early 20th century. Most of them designed their space stations as huge rotating wheels with the inhabitants residing inside of the rim, thus avoiding the sensation of being weightless all of the time. As mentioned previously, that is a very expensive way to do things compared to attaching a series of modules to each other and living in microgravity as on the ISS.

Nevertheless, it must be done, and NASA almost concedes this on a number of their recent planning charts that are available on line. There are other charts, however, that align with Dr. Robinson's comments referenced earlier. They show pharmaceutical and exercise countermeasures being required for extended stays on the Martian surface. That is an interesting conclusion to draw, seeing as how there are no data concerning how the human body (or that of any other mammal) will react to 0.38g!

The Dirt on Centripetal Acceleration

Unfortunately, the force exerted on a body inside of a rotating container does not generate isolated linear acceleration. Other forces and associated dynamics are being generated simultaneously. To explain this, let us begin with everything at rest and pretend that we are floating inside of a box. As shown in Figure 7, the box begins to swing about a central axis, but you have mass and nothing is touching you, so you do not move. ("An object at rest tends to stay at rest...") Forget the air since it would not generate enough force to propel you.

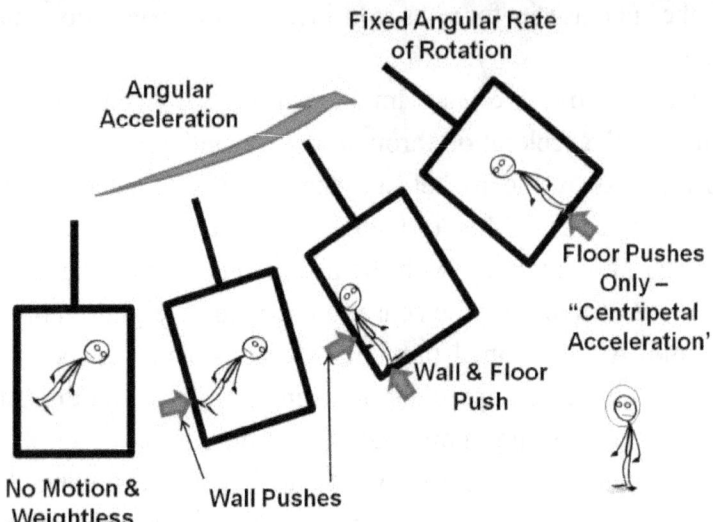

Figure 7. Using centripetal acceleration to create "artificial gravity"

The wall reaches you and begins to push you along with it as it continues to accelerate. However, because the box is going in a circle, the wall and floor move "up" relative to your body. This continues until your feet reach the floor, at which time the floor begins to push on you in addition to the wall. Suddenly you feel as if you have weight. As long as the angular rate (rotation rate) of the box is changing, you will experience pressure from the wall and the floor. When the box and arm reach a steady rotation rate (no more acceleration), the wall will cease to push on you but the floor will continue to exert the force required to keep you inside of the box.

Since there is a force, there must be an acceleration (remember a=F/m?) and that is what we call "centripetal acceleration." If you wish to get your brain really fried, type "does centripetal acceleration really exist?" into your search engine. (However, do not do it until you finish the book; I do not want to lose you.)

Stand perfectly still in the box when the angular rate is constant and you have "pseudo-gravity," more commonly called artificial gravity. Theoretically your various organs and bodily functions will not know the difference between what you are sensing and being in a gravity field. However, if anything moves you will observe something labeled "Coriolis Effects" and illusions of being in a simple gravity field disappear quickly. For example, if you drop a rubber ball it does not "fall" straight down. Its path is curved on the way down and curved on the way up. If you face in the direction of rotation and throw the ball just so, it will make an arc and come back to you.

The Coriolis Effects become most noticeable to the human body and brain when moving or rotating of the head occurs in the presence of whole body rotation. The nausea and related effects are due to a conflict between what the inner ear, the body, and the eyes are telling the brain. These effects worsen as the rotating arm gets shorter and the rotation rate increases.

There are numerous papers in the artificial gravity literature that propose so-called "comfort zones," none of which are necessarily optimized for Mars gravity. A search of the literature on-line will lead to plenty of references; many of the latest are included in Appendix B: Bibliography. However, I have found the following published paper to be the most helpful in bringing order to the chaos. "Artificial Gravity Visualization, Empathy, and Design" Theodore W. Hall, AIAA 2006-7321, Space 2006, 19-21 September 2006.

Figure 8 is adapted from Hall's paper. His work blends all of the disparate recommendations into one nomograph with which to select spin rate and rotator arm length. The reference paper is highly recommended for anyone interested in the ancillary phenomena of rotation-generated artificial gravity and the effects on proper design of human spaces.

As an aside, for *The Earth-Mars Chronicles* I selected a 340-meter (1115-feet) rotating arm length and a spin rate of one revolution per minute for the Mars transports and the Earth-Mars Station. Why? Because I took the easy way out. At that length and rotation rate, I did not have to keep track in my narrative of the Coriolis Effects because they become greatly reduced with long rotation arm length and low rotation rates.

You will see in a later chapter that the test, evaluation, and training station I propose has a rotation rate of two rpm with an 85-meter (279-feet) swing arm. From a practical design point of view, this works out to be a very reasonable compromise. All of the reports I found on experiments with swing arms and rotating rooms concluded that two revolutions per minute was easily adapted to in the presence of much shorter radii. Eighty-five meters should be a piece of cake.

The other advantage to this radius is that the equivalent of one Earth gravity can be generated at a rotation rate of 3.24 rpm, which is well below the 4 rpm used in the Joosten study referenced earlier. This combination of rotator arm length and rotation rate also falls into the comfort zone of Figure 8. Filling in data points to determine the shape of the curves in both Figure 7 and Figure 8 will be a simple matter of varying the rotation rate of a properly designed test and evaluation station.

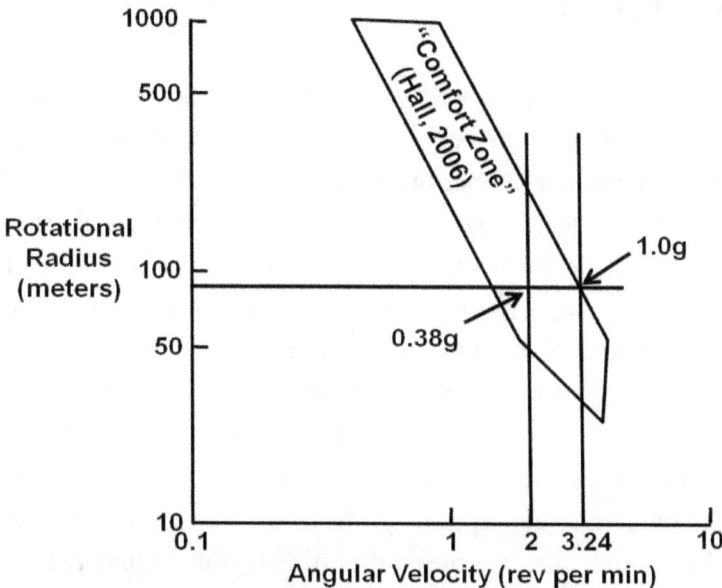

Figure 8. Hall's generally agreed to "comfort zone"

What Price Caution

No one of good conscience ever wishes a space mission to fail, but it happens. Every time it does we find out that the warning signs were there, but it was going to cost too much or take too long to mitigate the problem, so someone chose to stick their head in the sand and proceed. Lives and treasure are lost that way. We must not allow this to happen on the way to Mars. Now does that mean that we should never attempt to go to Mars to explore and begin our settlements? Of course not! It means that lessons should be learned from history. The Challenger tragedy is probably the best example in the modern age of space. "Err on the side of caution" must be the mantra, not "no mission until everyone agrees that the risk is zero."

Risk in any challenging endeavor is an unpleasant fact; all we can ever hope to do is reduce the probability of failure to the limit of our ability. An expedition to Mars elevates the challenge of alleviating risk to an entirely new level, one never experienced before. Unfortunately, this fact may have been obscured by the resounding success of the Mars rovers Spirit, Opportunity, and Curiosity. Thus far humans, the weakest link in the Mars expedition chain, have been nowhere near Mars.

This is not to say that considerable effort has not been expended on understanding human factors that may affect success. Many simulations of missions with small groups of people have been conducted, one even lasting over 500 days. The volunteers live in cramped quarters and communication time delays of several minutes are faithfully reproduced, along with the extension of the length of a day by 35 minutes and 39 seconds in some cases. Typically, everything possible is done to make the trip and the stay on Mars as real as possible.

There are two things that cannot be simulated on Earth in any fashion. One is true isolation with no hope of rescue, and the other is the attraction of gravity. The first lends itself to evaluation by corollary experiences and testing. However, the total lack of information and experience could make the effects of living in 0.38g the most sinister danger the members of the expedition may face. As discussed in an earlier chapter, many side effects from living in microgravity have been documented and attempts have been made to mitigate them, with varying degrees of success. No one knows what will happen to human beings who are exposed to 0.38g for extended periods. Success (or failure) with pharma and exercise solutions at microgravity cannot be extrapolated to any gravity level, including that of Mars.

One disturbing trend recently has been comments to the effect that, except for vision issues, all of the astronauts have made a full recovery after returning to Earth gravity. Time required for recovery has varied with both the length of exposure to microgravity and with the individual. Now Astronaut Kelly and Cosmonaut Kornienko are going to extend the experiment to one year, and for what? The answer is to "prove" that artificial gravity is not needed for humans travelling to and from Mars.

This is disturbing since it will prove nothing. It does not take 342 days to get to Mars, and they will not be alone on an alien planet 200 million kilometers from home, left to adapt to 0.38g such as they would be on a real expedition to Mars. Nothing will be known about whether the regimen of drugs and exercise is beneficial on the Red Planet. Using the precious volume and weight on the entry, descent, and landing vehicle to carry exercise equipment seems a bit ludicrous, especially when no one knows whether it is needed or of any use. Even if such equipment were available, no one would know what exercises and durations were needed.

Another alternative to artificial gravity that has gotten considerable attention recently is using a short arm centrifuge to periodically expose the astronauts to Earth gravity forces or higher. This is sometimes referred to as "hyper-conditioning" when the centripetal force exceeds Earth normal gravity. Opinions are diverse on the efficacy of this approach.

One logical argument takes the position that a large portion of a 24-hour day would have to be devoted to "conditioning" in order to reap the benefits, which would be unacceptable. Regardless of the outcome of such experiments in microgravity, the questions regarding physiological response in Mars gravity will remain unanswered. One thing we can be sure of is that using drugs and massive exercise in an attempt to avoid the red area of Figure 3 over an extended or forever stay on Mars is not acceptable. It is also likely that to do so without any hard data will be unacceptable to the astronauts.

Conducting in-situ experiments on Mars to determine the effects of 0.38g on fetus and child is obviously not acceptable. Only one facility in Earth orbit would be required to answer all of our questions.

Mars Close to Home

The hypo-gravity issues can no longer be overlooked or ignored. It will take several years, perhaps ten or more, after the test and evaluation facility is operational to obtain the answers needed concerning both exploration and settlement of Mars. Those experimental results may significantly influence plans for any trip to the Red Planet.

The literature on this subject is loaded with comments contending that pharmacology and exercise, though helpful in impeding specific manifestations of the problems, will never be totally effective countermeasures for the full spectrum of microgravity effects. However, there are no experimental data addressing the effectiveness of any pharma or physical attempt at mediation in a 0.38g environment. In fact, we do not know if such measures are required in the presence of Mars gravity or pseudo-gravity. Again, I emphasize that all of this discussion applies only to those extremely fit people who qualify as astronauts!

Russian scientists have some data suggesting that 0.3g *might* be sufficient to stave off some or all of the muscular degeneration that has been observed, but that can only give us hope. We must have data on multiple human beings living in 0.38g over an extended period in controlled conditions before committing lives to travel to and visit or live on Mars.

Simply put, we (the scientists, engineers, and authors) do not have any foundation for an expectation of happy and healthy astronauts, Settlers, and offspring on Mars. We have simply hoped for the best. This state of affairs must be rectified; it has persisted far too long.

So... what needs to be done? It does not require an intensive literature search to find the answer. The lack of sub-g data is now a warm topic. Blogs, op-eds, articles, and interviews are plentiful expressing concern over the wisdom of NASAs approach to addressing microgravity and partial Earth gravity. It appears that there could be broad support for a substantial program devoted to addressing all of the issues associated with the Martian environment. As several have proposed, one facility addressing those issues should be an environmental research station close to Earth.

Conceptually, such a facility could begin simply and then expand its capacity. When expanded to full scope, it is conceivable that it would be capable of emulating everything on Mars but the depth of the atmosphere and the dust devils. Every important issue – gravity effects, plant growth, human adaptation, dust storms, everything - could be addressed within the station for months or years at a time. Its best feature would be that it is *close to home*; a minor or major failure in design, fabrication, or function would be a test anomaly, not a fatal error.

Additional benefits of immeasurable value would also accrue. Engineers and technicians would gain experience in designing, fabricating, building, and operating a large rotating structure in space. Such experience and knowledge cannot be procured in any other fashion. They would be prepared, and they would prepare the next generation, to design the interplanetary ships that will have artificial gravity accommodations.

***

For over a century it has been widely accepted that a picture is worth a thousand words, so perhaps a rendering of a

Mars Test, Evaluation, and Simulation Station (TESS) concept is in order. However, before presenting you with a visual representation of a design concept, let us review what drives the design.

The following summarizes the problems, solutions, and benefits that dictated the requirements for the example concept design. *Addressing how well hardware, software, and astronauts meet those requirements by traveling through interplanetary space to Mars in order to determine the consequences must be deemed unacceptable.*

Known Problems and Challenges to Consider

1. Known microgravity health issues and lack of ANY data on:

a. Human responses, at any age, to hypogravity, especially 0.38g which cannot be suitably duplicated on Earth.

b. Human responses, at any age, to a very long-arm rotating environment at two to four rpm with a simulated gravity of 0.38g and 1.0g.

c. Viability of pregnancy, pre-birth development, infant growth, and adolescent development in a 0.38g environment (includes both mental and physical development).

2. Lack of capability to test equipment bound for Mars in a high fidelity simulation of conditions on the Red Planet, to include gravity.

3. Lack of capability to evaluate and train astronauts in a high fidelity simulation, with or without gravity, of a Mars Transit Vehicle.

4. Lack of capability to train astronauts in a high fidelity simulation of conditions on Mars, to include gravity.

5. Lack of knowledge on reconditioning required for astronauts returning to Earth from Mars.

6. No facilities or experience in Earth orbit to support design, logistics, transporting, assembly, testing, and operating of large complex assemblages such as interplanetary spacecraft and artificial gravity habitats.

Solution

Build a facility in orbit capable of simulating Mars and Earth gravity utilizing rotation; *design it to properly address problems 1 through 6*. Equip it with an environmental chamber to simulate the atmosphere, regolith, dust, and day-night cycle on Mars. Build the chamber with the capacity to house and thoroughly test the equipment to be landed on Mars. Incorporate the capability to integrate an actual Mars Transit Vehicle habitat for system testing and deep space travel training for astronauts.

Rationale

No other solution will meet the requirements.

Benefits

1. A carefully designed program of research executed on the station would provide answers to the fundamental questions surrounding astronaut health and viability during an exploratory trip to Mars and back.

2. A carefully designed program of research executed on the station would provide answers to the fundamental questions surrounding progeneration and long-term health prospects for generations born and living on Mars and in an artificial rotating environment.

3. Use of the station would substantially reduce risk by testing in a high fidelity environment the actual Mars transit and surface systems. This would include habitats, surface excursion pressure suits, rovers of various design, in space and surface robots, in-situ resource utilization (ISRU)

handling and processing equipment, environmental control and life support systems, deploying and retrieving of equipment, and most other functions to be exercised in space or on Mars.

4. Agricultural concepts and techniques could be tested with the correct simulated gravity, regolith, and other environmental factors.

5. Design, fabrication, construction, and operation of the orbiting test, evaluation, and training facility would build a foundation for future near-Earth and deep space mission planning and execution.

6. Purchasing water for various purposes and raw materials for regolith and ballast from asteroid or lunar miners would provide a basis for development of a true space-based industry. These same suppliers would one day be able to provide raw materials for use with so-called 3D printers and other fabrication techniques being used in orbit.

***

A derived requirements statement for the station provided the fundamental guidance required to crunch the numbers and gain insight into what a Mars TESS might look like. The following is the logic train that turned the notional concept above into an engineering concept.

Mission:

1.   Determine whether progeneration and maturing of humans in 0.38g is safe and acceptable through multiple generations.

2.   Determine both qualitatively and quantitatively the effects of long-term exposure of astronauts to the surface conditions on Mars, including those due to gravity.

3. Determine the suitability of Mars bound equipment including robots for use in the environment on the Martian surface.

4. Test and evaluate the Mars Transfer Habitat under the influence of microgravity, followed by characterization at simulated 0.38g and 1.0g in the presence of Coriolis Effects.

5. Provide a high fidelity training environment in microgravity and 0.38g for Mars mission astronauts and Settlers.

6. Evaluate and demonstrate techniques and procedures required to support assembly in orbit of interplanetary spacecraft.

7. Provide space operations coordination and mission control for "tugs," crew and resupply flights from Earth, lunar traffic, and others as appropriate.

8. Enhance education and promote international cooperation through interactive worldwide programming.

9. Cultivate cooperation of government and private enterprise by providing, on a non-interference basis, the first major tourist destination in space.

Objectives:

1. Observe and measure the effects on human subjects of long-term exposure to 0.38g while working and at rest. Identification and evaluation of countermeasures as required to alleviate any residual dysfunction or potential health hazard.

2. Define and evaluate regimens for preparation to return to Earth's gravity field including varying artificial gravity and incorporating exercise while on Mars and/or during the return trip.

3. Evaluate dust control strategies and mechanisms in a realistic environment, including 0.38g.

4. Extensively test and improve surface suit functionality in a realistic atmosphere, dust, and gravity environment.

5. Test and refine various mechanical techniques for fabrication, joining, assembling, erecting, inflating, etc. both in space and on the surface of Mars.

6. Test various approaches to fabrication including 3D additive technology and evaluate the utility of each in a realistic environment.

7. Execute a program of testing, preparation, and training that is very visible to anyone on Earth as an educational activity.

8. Test and refine the functionality of robots for use in transportation, joining, assembly, and other tasks while in space and in the equivalent of Mars gravity, atmosphere, and surface conditions.

9. Simulate and then conduct near-space and deep space mission operations from Earth orbit.

10. Simulate operations from the surface of Mars and in orbit about Mars.

11. Accommodate visitors on a noninterference basis.

Requirements:

Generate an engineering concept for a Mars Test, Evaluation, and Simulation Station (TESS) to be placed in low Earth orbit (LEO – orbital parameters TBD). The design shall meet the following requirements:

1. Generate an artificial gravity environment capable of being varied from microgravity to 1.0g with a rotation rate not to exceed four (4) revolutions per minute (rpm) at 1.0g. The design point shall be 0.38g at 2 rpm.

2. Provide a Mars Testing and Training Facility (TTF) at the 0.38g location that simulates conditions on Mars complete with 20 centimeters (nominal mean thickness) of simulated regolith and dust. The chamber shall be of sufficient size to install and test the following:

a. Single and multiple Mars habitats of various configurations, sizes, and functions

b. Multiple surface rovers of various designs and sizes

c. The lower portion of a Mars Ascent Vehicle (MAV)

d. Multiple robots engaged in various simultaneous functions

In addition, the TTF simulated gravity test area shall be incrementally expandable to a factor of five or more modules.

3. Provide a simulated Mars atmosphere including dust storms in the test chamber that is adjustable to be equivalent to that on the surface at any location on the planet. The chamber shall also be capable of functioning with hard vacuum inside and out and at a maximum of 2/3 Earth standard air pressure (675 mbar, 9.8 lb/sq in) inside.

4. Two Bigelow Aerospace B330 units shall be provided to support initial operational capability. They shall be configured to house a TESS control center, a TTF operations center, Mars and station crew living accommodations, laboratories, and a simulation center for Mars crew training. Total accommodations for crews TBD.

5. Accommodations shall be provided for attachment of a Mars Transit Habitat (interface specifications TBD) to be located separate from the TTF specified above.

6. Attachments and utilities shall also be provided for up to eight (8) Bigelow Aerospace B330 modules.

7. Additional B330 modules, up to a maximum of eight units, shall be attached as the environmental test chamber is expanded. These modules shall house accommodations for additional crew and trainees as well as for visitors (numbers TBD). More laboratories shall be installed as well as expanded control facilities for managing construction projects near the TESS.

8.   The TESS shall provide sufficient microgravity volume in the rotation hub to allow the addition of laboratories and training modules. Size and shape of these modules are TBD.

9.   An inertially fixed docking port shall be provided at the station hub compatible with international standards for docking interfaces in space.

10. Pressurized tunnels shall be provided allowing access to all modules, the TTF, and the hub of the station.

11. Power to the station shall be provided by either solar cell arrays or a small thermonuclear reactor (TBD).

The TESS design concept presented in the next chapter meets these requirements as discussed in Appendix A.

Test, Evaluation, and Simulation Station (TESS)

This version of a Test, Evaluation, and Simulation Station (TESS) is one step away from what is generally referred to as a back-of-the-envelope concept sketch. The intent of the exercise was to make sure that a rotating station of reasonable size and mass could be designed to meet the hypothetical requirements listed above. The answer is yes – depending on what the reader considers reasonable.

TESS is large by all space standards but not a "monster" when compared to the International Space Station or many robust Earth-bound engineering projects. TESS is also massive, but again by space standards, not relative to Earthly

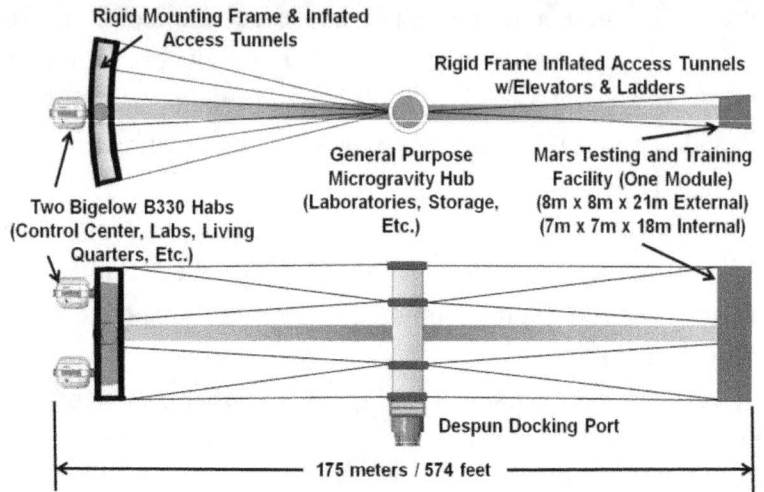

Figure 9. Phase 1 Test, Evaluation, and Simulation Station (TESS)

engineering feats.

The following TESS concept meets all of the requirements spelled out above. Gory and boring details of each response to specification are contained in Appendix A. Emphasis here is on form following function, so changes to the specs will change the results, but the intent was a simple examination of feasibility and practicality. This design concept shows that such a station could be built to satisfy its intended use.

TESS Dimensions and Mass Estimates

The various functional areas of the Test, Evaluation, and Simulation Station were designed to allow modular growth. Figure 9 illustrates Phase 1 consisting of one Testing and Training Facility (TTF) module on one end of the station and two Bigelow Aerospace B330 modules attached to a mounting frame on the opposing end. The mounting framework is attached to the hub structure via cables and the pressurized passageway. The framework provides attachment docks and restraints for a Mars Transit Vehicle Habitat and up to eight B330s.

Contained within the mounting framework is a pressurized access way connecting modules to each other and to the pressurized passageway housed inside of the rotation arm. A second rotation arm attaches to the hub at one end and the TTF module at the other. Within each pressurized cylindrical rotation arm is a rectangular elevator that ascends and descends between two ladders, one located rotation forward, and one attached rotation aft. As Hall points out in his paper referenced earlier, a single ladder cannot safely fulfill both up and down climbing functions due to Coriolis Effects.

Overall length of the station is about 175 meters, perhaps a little more. That puts the 0.38g level at the floor of the TTF and at the bottom floor of both the MTV Habitat and the B330s. The effective rotation length of 85 meters and 2-rpm rotation rate were specified based on consideration of minimizing Coriolis Effects and the acceleration gradient between head and foot of the astronauts. For an astronaut 183 cm (6 ft) tall the acceleration at the head would be 0.372g, creating a two percent gradient that would not be noticeable.

It is expected that a priority for the station during Phase 1 would be evaluation of the physiological effects of varying artificial gravity levels, starting with 0.38g, the acceleration of gravity on the surface of Mars. The station would also be capable of providing other hypogravity levels by adjusting the rotation rate. Earth's 1g can be simulated at a rotation rate of 3.24 rpm. Although on the edge, this rotation rate is still within Hall's "Comfort Zone" as illustrated on Figure 8.

Various forms of hardware test and evaluation would be performed in the TTF parallel to the research being done on physiological effects of 0.38g. These tests would achieve two objectives. Testing of the hardware in a high fidelity Mars simulation environment would be one, and evaluation of the TTF functionality would be the other. Undoubtedly, improvements would be incorporated into the TTF.

Loading or unloading the TTF could be done by ceasing station rotation and opening the "end" of one of the outboard modules. Strategic placement of load handling arms inside and outside of the TTF such as used on the ISS would provide handling and placement of large test items.

Phase 2 for the station is shown in Figure 10. Two TTF modules have been added to increase the test and evaluation floor area and a Mars Transit Vehicle Habitat has been placed between the two B330s. The Habitat could be a special test article or it could be actual flight hardware. It seems likely that the MTV Habitat would undergo a complete round trip simulation in real time.

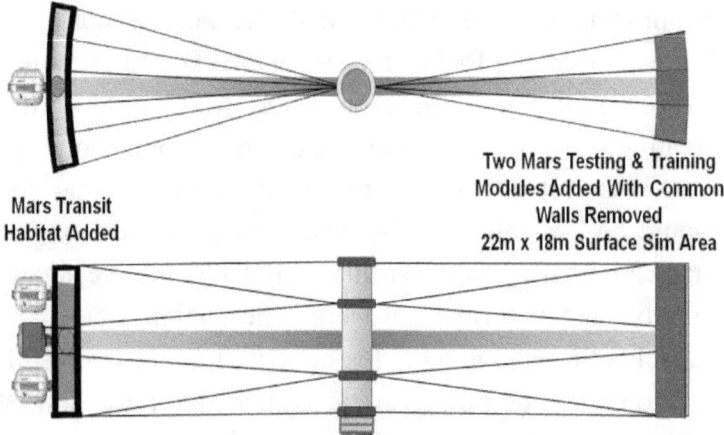

Two Mars Testing & Training Modules Added With Common Walls Removed
22m x 18m Surface Sim Area

Mars Transit Habitat Added

Figure 10. Phase 2 Test, Evaluation, and Simulation Station

Testing of the surface habitat would be another priority, along with any support modules devoted to in-situ resource processing. Surface suits, rovers, and other support equipment would also be tested and operated in the simulated Mars environment. If required, two more TTF modules could be added along with up to six additional B330s. This Phase 3 configuration is shown on Figure 11.

The following mass estimates are based on a combination of Johnson Space Center estimating factors, similarity to actual hardware, web site vendor data, and good old fashion calculations. Note for comparison that the ISS "finished" mass is estimated to be 420 metric tons (MT) or 925,000 pounds Earth weight. Also, note that no mass estimates for power generation equipment are included. (All pounds are Earth weight equivalents.)

Phase 1 Mass = 150 MT (330,000 lb)

Including 38 MT (83,600 lb) of simulated regolith

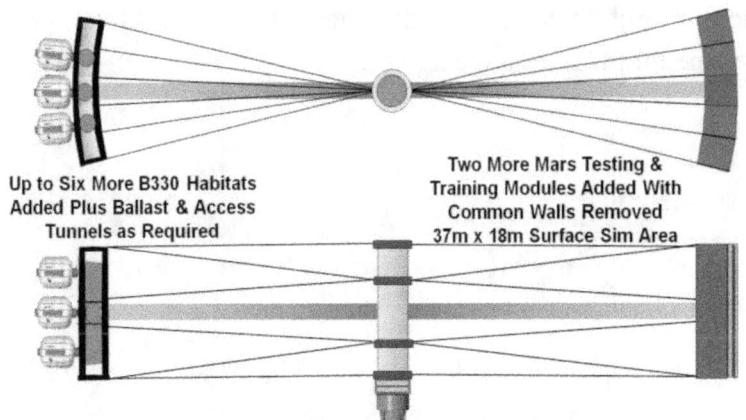

Figure 11. Phase 3 Test, Evaluation, and Simulation Station

Phase 2 Mass = 378 MT (830,600 lb)

Including 114 MT (250,800 lb) of simulated regolith

Phase 3 Mass = 610 MT (1,342,000 lb)

Including 190 MT (418,000 lb) of simulated regolith

Even though the simulated regolith is only 20 centimeters (7.87 inches) thick, it still contributes substantially to the total mass due to having a specific mass of 1,520 kg/cubic meter. This in turn requires that the Phase 3 habitat end of the rotating structure carry an additional 72 MT of station hardware or dumb ballast to balance the load. Using a lower density ersatz regolith or none at all would dramatically reduce the total mass. However, that decision would require a close look at those test and evaluation or training objectives that would be compromised.

Figure 12 presents (roughly to scale) a cartoon of an inflatable habitat and an exploration rover installed in the TTF for extensive testing and training. No optimization of module size for the TTF was conducted and larger modules might be more useful than the 8x8x20 meter (26x26x66 feet) units assumed for this design exercise.

Figure 12. Cartoon of an inflatable habitat and exploration rover installed in TTF

Gateway To The Future

Figure 13 shows a comparison to scale between the USS Enterprise Aircraft Carrier, the International Space Station, and the Mars Test, Evaluation, and Simulation Station concept. Obviously, the Mars TESS would be a large space project. It is not, however, something that we cannot handle. After all, it is essential that we leave behind the outmoded Apollo thought process and forge a new paradigm for our expansion into the solar system.

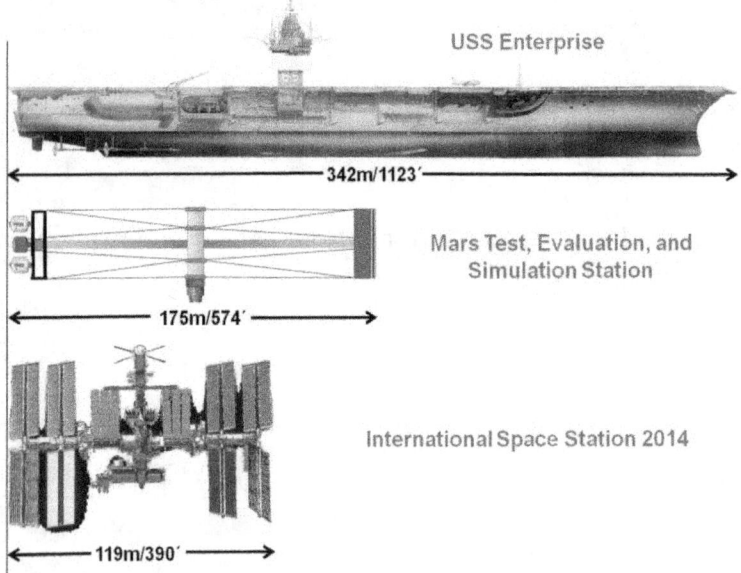

Figure 13. TESS compared to a large Earth project and a large space project

Based on his actions, it appears that Elon Musk understands this. Buzz Aldrin obviously does; he stated as much in his book. Many in NASA also understand that this paradigm shift must take place to make space faring practical. Building ever larger and more expensive rockets is not the answer, any more than building bigger and bigger covered wagons was the answer to opening up the western United States.

It is time to build something permanent in space, and the feat needs to be done as a joint venture of nations and commercial enterprise. An expandable Test, Evaluation, and Simulation Station would support government and privately funded Mars endeavors for decades to come. In addition, it would spur other applications such as the transition from throwaway satellites to those that can be refueled, repaired, and upgraded by providing a base of operations for commercial support companies.

A maintenance, repair, and upgrade facility for space tugs could be added to or replace TTF modules. It would be a relatively simple matter to stop the station rotation for pick up or launch using an arm such as the one used on the ISS. Obviously, any residual propellants or volatiles would be vented to space prior to bringing a tug on board the station.

The station would be a customer for water and other products derived from asteroid or lunar sources. It could also support other customers using those resources. Modules making up the Mars Testing and Training Facility could be converted to support manufacturing, or additional specialized modules could be added.

Simulated gravity could be adjusted to 1.0g or some other value if desired by changing the rotation rate of the station. Novel manufacturing processes combining microgravity, partial gravity, and full gravity would be possible. Lunar gravity and regolith conditions could also be simulated, allowing equipment, personnel, robots, rovers, and other items to be tested or trained prior to departing for the Moon.

Risks associated with establishing a settlement on Mars could be reduced by constructing the TESS large enough to accommodate a complete small settlement including a farm. The viability of raising multiple generations of crops on Mars having sufficient quality and quantity while maintaining all of the required balances (atmosphere, moisture, temperature, illumination, etc.) is debatable. It will remain so until a simulation of sufficient fidelity is conducted that demonstrates multiple generations and a breadth of product and nutrition. Once again, waiting until lives depend on the results on Mars is unacceptable. It also seems illogical to go all the way to Mars to find out how well or how poorly we are doing with our farming!

The dimensions chosen for the individual TTF modules were driven by the desire to manage launch cost by fitting all of the pieces for one unit into the large SpaceX faring. I have no idea what physical design factors such as weight, balance, moments of inertia, or stable rotation would become the limiting factor on how large the TTF could be.

Adding more modules to the TTF until it closed on both ends of the B330 and Habitat mounting frame would result in approximately 9,000 square meters (~97,000 square feet) of "floor space" to support expanded operations. If feasible, expanding the "width" of each module would also increase floor space. Each TTF module could be isolated from one or both neighbors, adding the flexibility to support multiple applications simultaneously.

All of that is well and good from the standpoint of supporting exploration of the solar system and promoting access to the minerals and other resources space has to offer. A much greater contribution will be the determination of whether human civilization will expand beyond the confines of the Earth's surface. Humanity will face a daunting realization if evolution in any gravity or pseudo-gravity field other than that of Earth proves totally unacceptable.

If no family unit of any type will ever live anywhere but Earth, an age-old question will be answered. Humanity has no place in the universe and will forever more be represented by nothing more than billions of sentient beings on a tiny blue and white ball, endlessly bickering over water and dirt.

Appendix A: Mars TESS Design Details

The following descriptions address the requirements in the order presented above. A "Concept Response" is provided for each one describing how the elements of the illustrations meet the selected requirement.

***

1.   Generate an artificial gravity environment capable of being varied from microgravity to 1.0g with a rotation rate not to exceed four (4) revolutions per minute (rpm) at 1.0g. The design point shall be 0.38g at 2 rpm.

Concept Response:

Rotating arm length must be 85 meters in order to satisfy the standard rotational relationship, i.e. acceleration = (rotation rate in radians/sec)^2 X (arm length in meters). You are spared the math.

At that length, it would be possible to produce a centripetal acceleration equivalent to one Earth standard gravity at a rotation rate of 3.24 rpm. Both of these design points fall within the "Comfort Zone" of Figure 8.

***

2.   Provide a Mars Testing and Training Facility (TTF) at the 0.38g location that imitates conditions on Mars complete with 20 centimeters of simulated regolith and dust. The chamber shall be of sufficient size to install and test the following:

a.   Single and multiple Mars habitats of various configurations, sizes, and functions

b. Multiple surface rovers of various sizes

c. The lower portion of a Mars Ascent Vehicle (MAV)
d. Multiple robots engaged in various simultaneous functions
In addition, the TTF simulated gravity test area shall be incrementally expandable to a factor of five or more modules.

Concept Response:

Various tradeoffs led to a "starter" chamber with a test area (or "floor") of 7 meters by 18 meters (23 feet by 59 feet). The surface area for test articles would be 126 sq m (1356 sq ft). Height (internal) was also set at 7 meters (23 ft) to enhance commonality of components between sidewalls and end caps. This yields a volume of approximately 882 cubic meters (31,124 cubic feet). The floor surface area is sufficient to house two habitats or one MAV but provides only limited room for external activities. Expanding the TTF is accomplished by moving the sidewalls out and adding floor, ceiling, and end caps.

***

3.   Provide an atmosphere in the test chamber that is adjustable to be equivalent to that on the surface of Mars at any location on the planet, including simulated dust storms. The chamber shall also be capable of functioning with hard vacuum inside and out and at a maximum of 2/3 Earth standard air pressure (675 mbar, 9.8 lb/sq in) inside.

Concept Response:

This requirement is not difficult to meet with simple panels designed for assembly in space and had no impact on the generous mass estimate. An "inflatable" chamber was not considered but should be if any further work is done on this concept. Some ingenuity would be required with regards to fan placement and dust injection and circulation to compensate for Coriolis Effects.

***

4.   Two B330 units shall be provided to support initial operational capability. They shall be configured to house a TESS control center, a TTF operations center, Mars and station crew living accommodations, laboratories, and a simulation center for Mars crew training. Total crew accommodations to be provided are TBD.

Concept Response:

Figure 9 shows the proposed location for the two B330 modules.

***

5.   Accommodations shall be provided for attachment of a Mars Transit Habitat (interface specifications TBD) to be located separate from the TTF specified above.

Concept Response:

Figure 10 shows the location of the habitat at the end opposite the TTF and flanked above and below by Bigelow Aerospace B330 modules. The habitat is the same dimensions and mass as the one used in the 2005 NASA study *Preliminary Assessment of Artificial Gravity Impacts to Deep-Space Vehicle Design*, Document No. EX-02-50. The hab module is also located at the 85-meter distance from the axis of rotation in order to impose the same simulated gravity effects as those in the TTF.

***

6.   Attachments and utilities shall also be provided for up to eight (8) Bigelow Aerospace B330 modules.

Concept Response:

The mounting framework and the primary pressurized connecting tunnel are illustrated in Figure 11.

***

7. Additional B330 modules, up to a maximum of eight units, shall be attached as the environmental test chamber is expanded. These modules shall house accommodations for additional crew and trainees as well as for visitors (numbers TBD). More laboratories shall be installed as well as expanded control facilities for managing construction projects near the TESS.

Design Response:

Figure 12 illustrates the Stage 3 TESS with five modules forming the TTF and eight B330 modules installed.

***

8. The TESS shall provide sufficient microgravity volume to allow the addition of laboratories and training modules. Size and shape of these modules are TBD.

Concept Response:

The hub of the facility shown in the figures is sized to allow volume for laboratories, training facilities, passageways, and stowage. If a despun facility is desired, it will be necessary to install mechanical or magnetic devices to do so.

***

9. An inertially fixed docking port shall be provided at the station hub compatible with international standards for docking interfaces in space.

Concept Response:

A despun docking port is illustrated in all three figures.

***

10. Pressurized tunnels shall be provided allowing access to all modules, the TTF, and the hub of the station.

Concept Response:

Passageways are visible in the structures used for module attachment. The two spokes extending from the hub are sized to accommodate an elevator plus an up and down ladder on opposite sides. Coriolis Effects dictate the requirement for separate ladders.

***

11. Power to the station shall be provided by either solar cell arrays or a small thermonuclear reactor (TBD).

Concept Response:
None required until power source specifications are available. Design details for attachment to the station would also depend on the orbit selected if solar power is utilized.

Appendix B: Bibliography (Limited to Most Apropos)

2014 International Workshop on Research and Operational Considerations for Artificial Gravity Countermeasures, Ames Research Center, February 19 – 20, 2014, NASA/TM-2014-217394,
Twenty-Three Papers

Artificial gravity as a countermeasure in long-duration space flight, Lackner, J. R., DiZio, P., Journal of Neurological Sciences, 2000 Oct. 15; 62(2):169-76

Final Report: Artificial Gravity As A Tool In Biology &Medicine, Study Group 2.2, International Academy of Astronautics, 2007 June 5

Study on Artificial Gravity Research To Enable Human Space Exploration, International Academy of Astronautics, September 2009, ISBN 978-2-917761-04-5

Variable Gravity Laboratory for Deep-Space Crewed Missions Research and Demonstration,
Gary P. Noyes, Oceaneering Space Systems, AIAA-2011-5130

The Architecture of Artificial Gravity: Mathematical Musings On Designing For Life And Motion In A Centripetally Accelerated Environment, T. Hall, Univ. of Mich., Proceedings of the Tenth Princeton/AIAA/SSI Conference, May 15-18, 1991

Design Concepts for a Manned Artificial Gravity Research Facility, J A Carroll, Tether Applications, Inc., Space Manufacturing 14: Critical Technologies for Space Settlement - Space Studies Institute October 29-31, 2010

Artificial Gravity Visualization, Empathy, and Design, T W Hall, Univ. of Mich., Space 2006, 19-21 Sep 2006, AIAA 2006-7321

A Tether-Based Variable-Gravity Research Facility Concept, Kirk Sorenson, NASA Marshall Space Flight Center, 30 Nov 2005.

Another Go-Around: Revisiting the Case for Space-Based Centrifuges, L R Young, et al, Gravitational and Space Biology, Volume 25 (1), Sept 2011

Design Concepts for a Manned Artificial Gravity Research Facility, J. Carroll, Tether Applications, Inc., 2010 IAF Congress, Prague, 27 Sep 2010

NASA Strategic Plan 2014
Full Strategic Plan, Performance Plan, and Budget: http://nasa.gov/news/budget/index.htm
Agency Priority Goals: http://goals.performance.gov/agency/nasa
Cross-Agency Priority Goals: http://goals.perfor mance.gov/goals_2013

Second Mars Affordability and Sustainability Workshop, 14-16 Oct 2014, The Keck Institute for Space Studies; The California Institute of Technology; Hosted by the NASA Jet Propulsion Laboratory, Organized by Explore Mars, Inc. and the American Astronautical Society

Preliminary Assessment of Artificial Gravity Impacts to Deep-Space Vehicle Design, B. Kent Joosten, Johnson Space Center, 11 May 2005, Document No. EX-02-50

Summary of the Final Report, Mars Program Planning Group, 25 Sep 2012

Space Launch System and NASA's Journey to Mars, Todd May, SLS Program Manager, Mars Society 2014 Annual Meeting, 14 August 2014

In-Space Propulsion System Road Map, Technology Area 02, M. Meyer, L. Johnson Co-chairs, April 2012

Advanced Exploration Systems, J. Crusan, Director Advanced Exploration Systems, NASA Advisory Council – Exploration Committee, 10 December 2013

NUCLEAR THERMAL ROCKET/VEHICLE CHARACTERISTICS AND SENSITIVITY TRADES FOR NASA's MARS DESIGN REFERENCE ARCHITECTURE (DRA) 5.0 STUDY, S. K. Borowski, et al, Proceedings of Nuclear and Emerging Technologies for Space 2009, Paper 203599, Atlanta, GA, June 14-19, 2009

Author Biography

Gerald W. Driggers has BS and MS degrees in Aerospace Engineering from Auburn University. His 45-year career in science and engineering spanned many disciplines and gave him an uncommon depth of understanding in things both theoretical and practical. Military experience varied from launching ballistic payloads and satellites to battlefield Civil Engineering. In civilian life, his endeavors included design studies in space colonization and industrialization, original research to improve pollution control, groundbreaking developments in missile defense technology, and advancements in hardware-in-the-loop testing. In his leisure time, he mastered photography, automotive and outboard engine rebuilding, boat repair and maintenance, seamanship (including offshore), home restoration, and the art of raising two wonderful children. Gerald is truly a man of many trades and a master of several. Through it all, the exploration of the solar system has been his passion since age eight. Pinnacles in his professional life include time spent with Dr. von Braun and his team members, being presented an award by Arthur C. Clarke, conversations with Robert Heinlein, the friendship of Harry Stine, and his close relationship with Gerard and Tasha O'Neill. Gerald now writes about his passion; that is, when wonderful wife Carol or Wilson the Cat do not have chores for him.

Works by Gerald W. Driggers
Fiction:
Martian Sniper: A War for Cousins **Series**
The Earth-Mars Chronicles **Series**
Butterscotch Dawn: A Mars Exploration Adventure
Non-Fiction:
Mars Close to Home: The Case For a Mars Simulation in Earth Orbit
Gravity: How This Mysterious Force Will Dictate The Future of Humanity
The Forbidden: Why Families May Never Live Anywhere But Earth

Visit: www.earth-marspublishing.com
        www.earth-mars.com
        Earth-Mars on Facebook